A DIFFERENT
KIND OF ANIMAL

The University Center for Human Values Series

Stephen Macedo, Editor

A list of titles in this series appears at the back of the book.

A DIFFERENT
KIND OF ANIMAL

How Culture Transformed Our Species

ROBERT BOYD

PRINCETON UNIVERSITY PRESS
PRINCETON AND OXFORD

Copyright © 2018 by Princeton University Press

Published by Princeton University Press, 41 William Street,
Princeton, New Jersey 08540

In the United Kingdom: Princeton University Press,
6 Oxford Street, Woodstock, Oxfordshire OX20 1TR

press.princeton.edu

Jacket art: Archer, cave painting, Tassili n'Ajjer
(UNESCO World Heritage List, 1982), Algeria / De Agostini
Picture Library / M. Fantin / Bridgeman Images

Library of Congress Cataloging-in-Publication Data

Names: Boyd, Robert (Professor of cultural evolution), author.

Title: A different kind of animal : how culture transformed
our species / Robert Boyd.

Description: Princeton : Princeton University Press, [2018] |
Series: The University Center for human values series |
Includes bibliographical references and index.

Identifiers: LCCN 2017008706 | ISBN 9780691177731
(hardcover : alk. paper)

Subjects: LCSH: Social evolution. | Human evolution.

Classification: LCC GN360 .B685 2018 | DDC 303.4—dc23
LC record available at https://lccn.loc.gov/2017008706

British Library Cataloging-in-Publication Data is available

This book has been composed in Sabon

Printed on acid-free paper. ∞

Printed in the United States of America

1 3 5 7 9 10 8 6 4 2

Contents

ACKNOWLEDGMENTS

These lectures have grown out of a forty-year-long collaboration with my friend and mentor Pete Richerson. Pete is both a polymath whose knowledge of the social and biological worlds exceeds that of anybody I have ever met by a mile and an exceptionally insightful and creative thinker. Kind, practical, and generous—there is nobody better than Pete. I have also learned much from my former students, especially Joe Henrich, who took ideas that Pete and I developed and made them into an empirical research program par excellence. He's also pretty good in the "idea development" area himself. I'm glad at my age I don't have to measure up to the standard Joe sets. If you want to learn more about cultural evolution, Joe's recent book, *The Secret of Our Success*, is the best place to start. I have benefited from collaborations with a number of other terrific students including Michelle Kline, Richard McElreath, Cristina Moya, and Karthik Panchanathan. Recently, I have had the good fortune to collaborate with my colleague Sarah Mathew, an exceptionally clear thinker and a talented, productive field worker. Many of the ideas in the second lecture have their origin in Sarah's fertile brain. Five years ago I moved from UCLA to Arizona State University, and that brought me into contact with the researchers in the Institute of Human Origins (IHO). This has been a great boon. We secured funding from the John Templeton Foundation for a program of interdisciplinary research focused on the origins of human uniqueness. In the ongoing discussions, I have learned much from my IHO colleagues, especially Kim Hill, the greatest living ethnographer of hunter-gatherers and possessor of an exceptionally careful and creative intellect. Many of the ideas in this book arose from discussions with Kim. I have also benefited from Curtis Marean's encyclopedic knowledge of the archaeology of early modern humans and from Bill Kimbel's equally encyclopedic knowledge of Plio-Pleistocene hominins. I have had valuable interactions with IHO members Ian Gilby, Kevin Langergraber, Charles Perreault, Kaye Read, Gary Schwartz, and Anne Stone. My postdocs Maciek Chudek, Max Derex, and Hillary Lenfesty have been a source of ideas and helped keep me on my toes. A number of people read versions of this book in manuscript and provided helpful feedback, including Clark Barrett, Joe Henrich, Moshe Hoffman, Jillian Jordan, Alison Kalett, Ruth Mace, Stephen Macedo, Cristina Moya, H. Allen Orr, David Rand, Paul Seabright, and Kim Sterelny. Thanks to all. And special thanks to my longtime climbing partner and prose stylist extraordinaire John Wiley, whose edit greatly improved the writing. And last but way far from least, my spouse, Joan Silk, has provided a thirty-five-year in-home seminar on social evolution in primates, has been a clearheaded sounding board for my often half-baked ideas, and has provided good-humored editing that has greatly improved my often clunky prose, including that in the present book.

Some of the research discussed in these lectures was made possible through the support of a grant (ID: 48952) from the John Templeton Foundation to the Institute of Human Origins at Arizona State University. The opinions expressed in this publication are those of the authors and do not necessarily reflect the views of the John Templeton Foundation.

A DIFFERENT
KIND OF ANIMAL

INTRODUCTION

Stephen Macedo

What makes humans special? Is it, as many have argued, our superior intelligence that sets us apart from other species?

In the lectures and discussions that follow, Robert Boyd, a distinguished professor of human evolution and social change, refines the question and rejects the common answer. Putting aside the more familiar question of human uniqueness, Boyd asks why humans so exceed other species when it comes to broad indices of ecological success such as our ability to adapt to and thrive in such a wide variety of habitats across the globe. Ten thousand years ago, humans already occupied the entire globe except Antarctica and a few remote islands. No other species comes close. What explains our outlier status if not our "big brains"?

Humans adapt to a vast variety of changing environments not mainly by applying individual intelligence to solve problems, but rather via "cumulative cultural adaptation" and, over the longer term, Darwinian selection among cultures with different social norms and moral values. Not only are humans part of the natural world, argues Boyd, but *human culture* is part of the natural world. Culture makes us "a different kind of animal," and "culture is as much a part of human biology as our peculiar pelvis or the thick enamel that covers our molars."

With his many coauthors, especially Peter Richerson, Robert Boyd has for three decades pioneered an important approach to the study of human evolution that focuses on the population dynamics of culturally transmitted information. ("Cultural group selection" is a subset of this larger approach sometimes called "dual inheritance" or "cultural evolution.") That program is summarized, elaborated, and defended in the chapters that follow. Boyd's framework "provides a picture of human nature with powerful implications for how societies should be organized," and it "deserves to be much better known among scholars in the social sciences and humanities." So says the economist Paul Seabright in his contribution below, and I could not agree more. This volume furnishes a superb introduction for those with little or no background in evolutionary studies.

A few words, then, about the overall contours of this volume, which originated as the Tanner Lectures on Human Values at Princeton University in April 2016, organized under the auspices of the University Center for Human Values.

* * * * *

Robert Boyd marshals an astonishing range of scholarship, colorful vignettes, and anecdotes to argue that humans make use of insights and adaptations that we do not understand. We learn very often not by figuring out how things work but by imitating others who have locally useful "know-how." Boyd describes the conditions under which selection favors "a psychology that causes most people to adopt beliefs *just because* others hold those beliefs." Indeed, he argues that "even the simplest hunter-gatherer societies depend on tools and knowledge far too complex for individuals to acquire on their own." Culture is the storehouse of gradually accumulated, local, and typically tacit knowledge. "Cumulative cultural evolution" is the great and unique advantage of humans.

Not all of the consequences are positive: maladaptive ideas and false beliefs can also spread via blind imitation. But the dominant effect is that, as Seabright puts it, "thanks to a supremely flexible collective intelligence," we are "collectively smarter than any individuals in the population." Indeed, thanks to the power of imitative learning, "we do not need to be as intelligent individually as we are collectively."

In his second lecture, Boyd deploys his account of social learning and cumulative cultural evolution to illuminate how societies adapt to changing environments and develop ever more sophisticated tools and technology. Our ability to learn by imitation and our evolved trusting psychology are used to explain the centrality of social norms, and to explain why and how humans have for so long been "supercooperators." Even in foraging societies, the extent of human cooperation vastly exceeds that of any other species. Millennia of cumulative cultural evolution have helped create a vast "worldwide web of specialization and exchange." Humans are unique in that "people cooperate in large groups of almost unrelated individuals to provide public goods."

Boyd poses the puzzle thus: "Everywhere else in nature, large-scale cooperation is explained by kinship, but in humans it is not." So, "how could natural selection favor changes in human psychology that led to cooperation among large numbers of unrelated people?"

Cooperation in large groups "requires systems of norms enforced by sanctions." In larger and more complex societies, cooperation and the provision of public goods depend crucially on coercive sanctioning by third parties: institutions such as police and courts.

Cooperative social norms can take a great variety of forms, and societies have evolved a wide range of different moral codes: rules for marriage and inheritance, for example, and various political systems. Differing societies and cultural groups compete on the basis of these differing codes, which vary in their capacity to survive in a changing environment and prevail in competition with other societies. Christianity may have prevailed over paganism in the Roman Empire, for example, because whereas "paganism had weak traditions of mutual aid," "solicitous care of the sick" in Christian communities reduced mortality and increased well-being. Roman political institutions, on the other hand, have been adapted and persist.

Boyd acutely points out that evolutionary accounts of social life too often have a libertarian flavor, with society conceptualized as a "network of bilateral bargains among self-interested individuals and nepotistic families." Boyd rejects this picture and insists that even small-scale cooperation is "regulated by shared norms that are enforced by third-party sanctions," and that these provide "vital scaffolding in sustaining cooperation" among people in society.

One of the most notable features of Boyd's work generally and of the lectures that follow is, as I have noted, the breadth of the scholarly approaches and resources he draws upon. He and his students do fieldwork in Fiji and elsewhere, as cultural anthropologists, but he also engages in mathematical modeling, uses rational choice theorizing, and draws on any number of other scientific and social-scientific insights and methods. As he modestly puts it in his response in chapter 7, speaking of his long collaboration with Richerson, "Our research style was, and still is, to read widely in anthropology, psychology, and economics looking for promising empirical problems and then tackle those problems with theory derived mainly from population biology." Sounds simple, but few scholars approach his range and rigor.

"Cultural group selection" is not to be confused with moral or any other form of progress: Boyd's theory is social scientific and positive, not ethically normative. And yet no society can do without social norms and extensive cooperation, so there is no doubt that the ideas developed here are of great interest to anyone concerned with human nature and the social and institutional underpinnings of good and just societies.

Boyd concludes by reiterating his core thesis that "the evolution of cultural adaptation" was the "essential ingredient" in our "ecological success and our ability to cooperate." Humans are outliers in the natural world because "no other creature is able to create so many different local adaptations" that are "beyond the inventive capacities of individuals."

* * * * *

Robert Boyd's two lectures are followed by four critical comments written by distinguished scholars from a variety of disciplines.

H. Allen Orr, a general evolutionary biologist who studies speciation and adaptation and also writes for wider audiences, raises two large and interesting questions about Boyd's model of cultural learning. He wonders, first, whether Boyd exaggerates the contrast between the "Big Brain" model, which emphasizes cognitive explanations for human success, and the imitative model that Boyd prefers. Successful imitation often requires considerable "neuronal firepower," argues Orr. In addition, Orr usefully describes the partial convergence of Boyd's view with that advanced by the well-known free-market economist and social theorist Friedrich Hayek. Hayek also emphasized that social success and progress depend on the use of tacit and dispersed local knowledge, culturally transmitted social norms and ethical mores, and institutions that are the product of social evolution. Orr wonders whether scientists and social scientists pay less attention to Hayek than they should because of Hayek's politics.

Kim Sterelny, a prolific philosopher of science and especially of evolutionary biology, endorses, like our other commentators, the main contours of Boyd's argument: humans are outliers in our capacity to adapt to many environments. But, like Orr and Seabright, Sterelny asks whether Boyd goes too far in reducing the role of "our distinctive human intelligence" in explaining our ecological adaptability. He at least partly defends the "library" or "Big Brain" model that Boyd argues against. Tacit, practical know-how is a form of knowledge. In addition, Sterelny argues that Boyd relies too heavily on a simple and "conformist" or "trusting social learning heuristic." As a final point, Sterelny wonders whether and how social learning has changed across "domains and across time."

Ruth Mace, who is, like Boyd, an anthropologist but whose approach is known as human evolutionary ecology, applauds Boyd's multidisciplinary approach to the study of human evolution, while stressing her own belief in the importance of empirical testing. She points out that many questions remain about how norms arise, why they vary, "how they are maintained, and how easily they change." In a more critical vein, Mace suggests that some of the behaviors that Boyd attributes to social norms and sanctions might better be explained based on individual benefits, including the decision to participate in warfare. She describes her own empirical research on intergroup conflict in Northern Ireland and raises the question of whether "competition and conflict between groups, such as interethnic warfare, leads to parochial altruism (that is, altruism directed only within the group)."

Paul Seabright, an unusually wide-ranging and influential economist, argues that there is a "darker dimension to what makes us human," which Boyd largely leaves aside. If we are "the most ecologically adaptable and massively cooperative species," argues Seabright, we are also "the most spectacularly and violently competitive, and the most deviously manipulative." Seabright contends that "a much larger part of the communication that takes place around norms in most societies is about individuals manipulating other individuals" than one would think from Boyd's examples.

* * * * *

In his reply at the conclusion of the volume, Robert Boyd expresses appreciation for the commentators' thoughtful disagreements, all of which "accept the value of trying to understand how culture evolved."

Boyd notes one broad point of contention, shared by Mace, Sterelny, and Seabright, which is that he does not "give people enough credit for making smart, well-informed decisions." Boyd stands his ground, arguing that individual choice matters but people's basic beliefs come from their social context.

With respect to the related comments by Orr and others, suffice it to say that Boyd expresses agreement that "cognitive abilities and cultural learning are mutually reinforcing."

Boyd ably defends his model against all four commentators and concludes by offering a pointed defense, against Seabright, of his own more optimistic view, closing with a most humorous observation that I will not spoil.

* * * * *

The program of research that Robert Boyd has pioneered along with Peter Richerson and their various coauthors, including Joseph Henrich and Sarah Mathew, provides a basis for reconsidering fundamental questions concerning human nature, social order, and human progress. The fascinating and astonishingly wide-ranging scholarship on display in the essays that follow is deeply suggestive for contemporary questions of institutional design and reform. Specific reform proposals await further scholarly inquiry which, I hope, this rich volume will help stimulate.

CHAPTER 1

NOT BY BRAINS ALONE

The Vital Role of Culture in Human Adaptation

STARVATION IN A LAND OF PLENTY

In 1860, the worthies of the city of Melbourne organized an expedition to explore Australia's interior, which was then unknown to white Australians. Their motives were mixed. Some hoped to find a route for a telegraph line that would connect Australia to Java, and then to the rest of the world. Others were motivated by rivalry with Adelaide, which had organized a similar, unsuccessful expedition a year earlier. They recruited Robert Burke, a dashing former military man, as the leader along with eighteen others, including William Wills, a scientist and cartographer. On August 20, the expedition set off with twenty-six camels, twenty-three horses, enough food for two years, and much Victorian impedimenta including oak dining furniture. The lead elements of the party, including both Burke and Wills, reached Cooper's Creek, a long string of ephemeral ponds about six hundred kilometers north of Melbourne, by November 11 and waited there for the rest of the party. By mid-December, the stragglers still had not arrived, and Burke had had enough. He, Wills, and two others, Charlie Gray and John King, set off hoping to reach the Gulf of Carpentaria and return in three months. Burke ordered the remainder of the party to wait for them at Cooper's Creek until March 15. Unfortunately, it took four arduous months to reach their goal and return, and Gray died along the way. When Burke, Wills, and King returned to Cooper's Creek in mid-April, they found the camp abandoned. They were exhausted and so low on provisions that they had no chance of covering the six hundred kilometers between Cooper's Creek and home. About this time, they were visited by a group of men from an Aboriginal group, the Yandruwandha, who, seeing their pitiable state, gave them six kilograms of fish. After a couple of weeks, they again encountered a group of Yandruwandha and accepted an invitation to their camp, where they were provided with more fish and cakes made from the seeds[1] of an aquatic fern called *nardoo*. The three white men liked the cakes and decided that if they were to survive, they must learn how to make nardoo flour. However, they had no idea what plant the seeds came from, and by then the Yandruwandha were nowhere to be found. After two weeks of desperate searching, Wills discovered the source of the nardoo seeds, and the three men began to collect and grind the seeds to make nardoo flour in quantity. However, despite having plenty of nardoo to eat, they gradually weakened, and by early July both Burke and Wills were dead. King was found by a Yandruwandha band that fed and cared for him for several months until a relief party arrived in September.[2]

So, why did Burke and Wills starve in what was a land of plenty for the Yandruwandha? The answer to this question holds the key to answering the much bigger question that is the focus of this essay: How did humans come to be such an exceptional species? Five million years ago our ancestors were just another, unremarkable ape. Today, our species dominates the world's biota. We occupy every part of the globe, we vastly outnumber every other terrestrial vertebrate, we process more energy than any other species, and we live in a wider range of social systems than any other creature. The key to this transformation is that people adapt culturally, gradually accumulating information crucial to survival. Central Australia was a land of plenty for the Yandruwandha because they were heirs to a rich trove of culturally transmitted knowledge about how to make a living there. As we will see, Burke and Wills died because they did not have access to this knowledge. In this essay, I make the case that our species has evolved the ability to adapt culturally, and this has, for better or worse, made us a different kind of animal.

ARE THE DIFFERENCES BETWEEN HUMANS AND OTHER ANIMALS REALLY THAT IMPORTANT?

Many of my colleagues from evolutionary biology don't think so. Of course, they would concede, people differ from other animals in lots of ways: we make much greater use of tools than any other creature; human language allows us to send a vastly larger range of signals than other animals; people have much bigger brains for their body size than most other mammals. But so what? Lots of animals have exceptional traits: elephants have long, flexible trunks, hummingbirds can hover, and indigo buntings navigate by watching the position of the stars. Moreover, we have a long history of thinking that we are different (and better) than other creatures, and many of our "unique" traits have turned out not to be unique at all. Biologists take great delight in disproving any claim that takes the form "only humans can do X." Toolmaking, language, farming, culture, teaching, and warfare have all suffered this fate.

How should we decide whether the differences between humans and other animals are important? Are humans really exceptional? Or are we just another unique species? I think that the best approach is to choose the same zoological criteria that we use to compare the ecological importance of other species, things like species range, biomass, and energy processing. If people really are different from other animals on these dimensions, then it's plausible that there is something anomalous about human adaptation that is worth investigating.

Let's start with species range. The geographical range of a species is the area that it occupies; its ecological range is the set of habitats it lives in. Both are useful because they give a rough measure of how adaptable a species is. All other things being equal, a species with a larger range must be able to function in a wider range of environments. Humans have the largest geographical and ecological range of any terrestrial vertebrate. Most terrestrial vertebrates occupy part of a continent and are limited to a modest range of habitats. Our closest relatives, the apes, are good examples. Orangutans and gibbons are limited to dense rain forest in Southeast Asia, gorillas and bonobos to moist forests in tropical Africa, and chimpanzees to forest and woodland in roughly the same parts of Africa. Big predators have the largest ranges; gray wolves occupy most of Eurasia and North America outside the tropics.[3] People live in every terrestrial habitat except Antarctica.

You might think that human expansion across the globe was a recent phenomenon made possible by agriculture and industrial production. But this is not the case. By the beginning of the Holocene, ten thousand years ago, hunter-gatherers had occupied every part of the globe except Antarctica and a few remote islands, and they lived in every kind of environment from the moist rain forests of Africa to the harsh deserts of Central Asia and the icy shores of the Arctic Ocean.

At this point, I'm sometimes asked about rodents: Don't they have a worldwide range too? And the answer is yes, they do. But rodents are an order that includes more than 2,200 species, almost all with much smaller ranges. Humans are different. One species, *Homo sapiens*, occupies virtually every terrestrial habitat on the planet. Norway rats are an exception, but one that proves the rule. Their range is almost as large as the range of humans (they don't survive in the Arctic). However, these creatures were limited to Central Asia until the Middle Ages and spread to the rest of the world by hitchhiking on human transport.[4] Norway rats and other human commensals (mites, helminths) live in similar human-created habitats everywhere.

Other standard zoological criteria tell the same story. Among vertebrates, human biomass[5] is exceeded only by that of our domesticates and is many times the biomass of all wild terrestrial vertebrate species combined.[6] The large human biomass is not just the result of agriculture and industrial production. It has been estimated that the carrying capacity for hunter-gatherers was about seventy million at the beginning of the Holocene.[7] This large biomass is notable because in many environments human foragers are top predators who hunt the largest animals in their habitat. Top predators are typically less numerous than their prey. For example, it has been estimated that at the beginning of the Holocene, lions—another large predator with a sizable range—numbered only about one million individuals.

The key to this ecological success is our ability to adapt to a wide range of different environments. Rodents live in virtually every habitat, but there are lots of different rodent species because rodents are specialists. Different species have different adaptations that allow them to succeed in particular environments.[8] Beavers have webbed feet and flat tails adapted to life in the water. Flying squirrels have large membranes that allow them to glide from tree to tree. Kangaroo rats have specialized kidneys that allow them to survive in deserts without having to drink water. Salt marsh mice can drink saltwater. And it's not just morphology.

Different species have different social behaviors adapted to their particular environments. Some are always monogamous, while others are highly promiscuous. Many species are solitary, but others, like beavers, prairie dogs, and naked mole rats, live in cooperative groups. In contrast, humans are generalists, able to adapt to a vast range of different environments and to develop local knowledge, specialized tools, and a wide variety of social arrangements.

Much human adaptation involves artifacts, but it is not just the ability to make artifacts that allows us to adapt to a wide range of environments. Many other animals make artifacts—bird nests, beaver dams, termite mounds, and the like—that play important roles in their lives. The technological sophistication of some of these artifacts rivals anything made by humans until the last few thousand years. The hanging nests of weaverbirds are beautifully designed and are at least as complex as the thatched dwellings made by many foragers. However, members of each weaverbird species construct a narrow range of nests.[9] What makes humans special is the ability to make many different kinds of artifacts that are appropriate in many different habitats. Humans, members of a single species, make different kinds of shelters in different places—houses made of snow, sod, stone, thatch, and wood that conform to many different designs, each well suited to the local environment.

Taken together, this evidence indicates that humans are exceptionally good at adapting to a wide range of environments. The obvious question is, why?

BRAINS ALONE AREN'T THE ANSWER

The standard, somewhat self-congratulatory answer to this question is that we are successful because we are very smart. Most treatments of human evolution take this account for granted. More complex tools, more sophisticated foraging, and symbolic behavior are all taken as indicators of increased cognitive ability.[10] For example, the archaeologist Lynn Wadley and her colleagues found that people living at Sibudu Cave on the coast of southern Africa about seventy thousand years ago used adhesives to fasten stone points to wooden shafts to make spears or arrows.[11] Their experiments demonstrate that these adhesives were made by mixing and heating acacia gum and red ochre, and that this procedure makes a better adhesive than plausible alternatives.

They conclude:

> Our experiments intimate that by at least 70 kya . . .
> these artisans were exceedingly skilled; they understood the properties of their adhesive ingredients, and
> they were able to manipulate them knowingly.[12]

There are two claims here. The first is that these artisans were extremely skilled. This follows directly from the fact that making and using adhesives is a complicated business. The second claim is that they were skilled *because* they were smart enough to understand the properties of these adhesives. As we shall see, the truth of the first claim does not imply the truth of the second.

Most people studying human evolution aren't very clear about exactly how intelligence translates into behavior and what intelligence really means in evolutionary terms. The simplest idea would be that we adapt to our environment just like other species, only better. Other species adapt to their environment in two ways. First, they change genetically. Individuals that live at higher latitudes are typically larger than members of the same species that live at lower latitudes, and this difference is adaptive because larger bodies are better at conserving heat. Natural selection favors differences in the distribution of genes influencing body size in populations living at high and low latitudes. Similarly, natural selection could favor certain behavioral adaptations (e.g., the ability to fling a spear) in one environment and other behavioral adaptations (e.g., the ability to ambush prey) in another environment. But natural selection is a relatively slow process, so genetic differences cannot explain why the people at Sibudu Cave used adhesives made from acacia gum while Aboriginal Australians used spinifex resin.

This leads us to the other form of adaptation: the flexible adjustment of behavior and morphology to the local environment during an animal's lifetime. In vertebrates, learning is especially important. Individuals learn where to find food, what to eat, and who will allow them to share a food patch, and they flexibly adjust their behavior in response to the local environment. This allows them to adapt rapidly to changing environments and enables a single species to occupy a diverse range of habitats.

It's important to understand that learning works mostly at the individual level in other species. Each individual learns what it needs to know on its own, not from others. It is true that lots of vertebrate species have simple traditions that are probably maintained by imitation and other social learning mechanisms. However, in every case, the traditions involve behaviors that individuals can and do acquire on their own.[13] Social information is helpful, but not essential. There is no doubt that individual learning can produce highly adaptive behavior and even complex artifacts without any social information. Weaverbirds, for example, weave beautiful nests without ever having seen one being made.[14] Young birds are not very good at nest building, but they improve with experience. Hand-raised woodpecker finches learn to use tools to extract grubs from under bark—exposure to competent adults is completely unnecessary.[15] Maybe people are just an extreme outlier of a similar process.

I'm going to try to convince you that this picture is wrong. Humans do not adapt as individuals like other animals. We are exceptionally smart, and this helps us adapt to a wide range of environments. But we are not nearly smart enough *as individuals* to solve the adaptive problems that confronted modern humans as they spread across the globe. The package of tools, foraging techniques, ecological knowledge, and social arrangements used by any group of foragers is far too complex for any individual to create. We are able to learn all the things we need to know in each of the many different environments in which we live only because we acquire information from others. We are much better at learning from others than other species are, and equally important, we are motivated to learn from others even when we do not understand why our models are doing what they are doing. This psychology allows human populations to accumulate pools of adaptive information that greatly exceed the inventive capacities of individuals. Cumulative cultural evolution is crucial for human adaptation. We humans would not be an exceptional species if we did not adapt culturally.

This brings us back to the ill-fated Burke-Wills expedition. Their story is just one example of what I call the "lost

European explorer experiment." It has been repeated many times during the past several centuries with similar results.[16] A small group of European explorers gets stranded in an unfamiliar habitat in which an indigenous population is flourishing. Despite desperate efforts and ample learning time, the lost explorers cannot figure out how to feed themselves, and they often die. If they do survive, it's frequently thanks to the hospitality of the indigenous population.

The men of the Burke-Wills expedition didn't die because they were stupid.[17] They died because they didn't have access to the culturally transmitted knowledge that allowed the Yandruwandha to survive around Cooper's Creek. Let's start with the nardoo. The white men would probably never have discovered nardoo on their own, but after seeing the Yandruwandha process the seeds into flour they were able to do the same. Yet they still died.

Wills knew something was wrong, and he wrote in his diary:

> I am weaker than ever although I have a good appetite and relish the nardu much but it seems to give us no nutriment & the birds here are so shy as not to be got at. Even if we got a good supply of fish I doubt whether we could do much work on them and the nardu alone, nothing now but the greatest good luck can now save any of us and as for myself, I may live four or five days.[18]

What Wills did not know is that nardoo contains an enzyme that degrades vitamin B1 in the intestine, and even though they were getting plenty of calories, they were suffering from beriberi, probably exacerbated by scurvy.[19] The Yandruwandha processed the nardoo by rinsing it in large amounts of water. Many plants contain toxins, and people living on plant-rich diets have detoxification methods to deal with them. Learning that nardoo is toxic and how to detoxify it is a hard problem, and because the Europeans did not have access to this culturally transmitted knowledge, nardoo was worse than useless—it was a poison.

Even so, there were other things to eat in this environment. Burke and Wills knew the nardoo wasn't working, and the Yandruwandha were catching enough fish to give strangers six kilograms. Why couldn't Burke and Wills catch fish? We don't know for sure, but here's one possibility. Many Aboriginal Australians used nets to catch fish in ponds, and the Diyari, another tribe that lived on Cooper's Creek a bit north of the Yandruwandha, did exactly that.[20] Again, Burke and Wills probably had the advantage of seeing the Yandruwandha use nets, so why not make their own? But this is another hard problem. To make nets, they would have had to learn how to make cordage from local materials. In this part of Australia, Aboriginal people used the bark and roots of verbine, a small bush with purple flowers, to make twine.[21] Once the plants were collected, the fiber had to be separated from the surrounding plant tissue. To do this, the whole plant was soaked in water for several days, and then the bark was stripped off, beaten, and manipulated until it became soft. Finally, it was dried and rolled into skeins. The fiber was spun by twirling it on the naked thigh and then twisting it into a two-ply line. Once the line was produced, you had to know how to weave an effective net. People in the Cooper's Creek region used two different techniques, each one good for certain kinds of fishing.

For the Yandruwandha, Cooper's Creek was a land of plenty because they had a rich trove of culturally transmitted knowledge about how to make a living there. A Yandruwandha "Natural History Handbook" would have run to hundreds of pages with sections on the habits of game, efficient hunting techniques, how to find water, how to process toxic ferns, yams, and cycads, and so on. Australian Aborigines are famous among archaeologists for the simplicity of their technology. Nonetheless, an "Instruction Manual for Technology" would have had to cover the manufacture and proper use of nets, baskets, houses, boomerangs, fire drills, spears and spear-throwers, poisons, adhesives, shields, bark boats, ground stone tools, and much more. Plus, as we'll see later, cooperation plays a crucial role in human subsistence. To become a competent Yandruwandha, you would also have needed to master "Social Policies and Procedures," "Grammar and Dictionary," and "Beliefs, Stories, and Songs," volumes of comparable length.

BUT WHY IS CULTURALLY EVOLVED KNOWLEDGE ANY GOOD?

So, fine, humans do not adapt in the same way as other animals. Instead of reinventing everything for ourselves, we can access a pool of adaptive cultural knowledge. But how exactly does this work? And how do people make sure that the information stored in the cultural pool is actually useful? There has been surprisingly little explicit analysis of this problem. But I think the common understanding goes like this. Innovation is hard. It not easy to learn that nardoo is poisonous, and even harder to figure out how to detoxify it. Determining whether behaviors are beneficial is relatively easy. So once an innovation occurs, people understand why it is beneficial, and it spreads. Remember Wadley's conclusions about the South African people who learned to make adhesives seventy thousand years ago: "These artisans were exceedingly skilled; they understood the properties of their adhesive ingredients, and they were able to manipulate them knowingly."[22]

In this view, culture is like a library. Libraries preserve useful knowledge created in the past. Librarians may determine which books are on the shelves, but the content of the books depends on the abilities of the authors, not the librarians. Highly skilled authors mean better books. In the same way, even though culture is necessary to preserve innovations, the usefulness of the innovations depends on the minds of the innovators. Without superior human cognitive abilities, there wouldn't be any complex adaptations to preserve.

This is how many people interested in human evolution think about culture. In a classic paper, John Tooby and the late Irven DeVore argue that humans have shifted into the "cognitive niche"—a way of life predicated on solving problems that other animals cannot solve.[23] We can invent new tools and foraging techniques needed in novel environments because we are better at causal reasoning and other forms of mental simulation than any other animal. The fact that we can learn from others means that everybody doesn't have to invent everything for themselves, and the cost of innovation can be spread over many individuals.[24]

This approach appeals to evolutionary thinkers because they think that explaining design in nature is a key problem.[25] By "design" they mean complex, improbable arrangements of matter that serve some function. For example, the eye is an organ that is clearly designed for seeing, and an eye works only if its components—the lens, iris, retina, and so on—have a very special structure, a structure that could not arise by chance. The same goes for complex behaviors like nardoo processing and net making. The only thing that can generate design in nature is natural selection and things like the human brain that have been designed by natural selection. Culture, according to this view, is a compendium of useful innovations that are preserved because individuals appreciate their value.

An alternative view, widespread in anthropology, sociology, and psychology, is that people adopt the beliefs and practices of those around them. These beliefs and practices may be sensible, but they don't have to be. Moreover, people may not have a clear understanding of why one belief or practice is preferable to alternative beliefs or practices. People come to believe the things they believe and behave the way they behave because that is what other people in their culture do.

So which picture is correct? Do individuals adopt beliefs and practices because they understand why they are beneficial? Or do they adopt beliefs and practices because that's what the people around them believe and do? And, if the latter, why are the beliefs and practices useful?

DO PEOPLE USUALLY KNOW WHAT
THEY ARE DOING?

For example, did the Yandruwandha understand why they processed nardoo? We will never know because the Yandruwandha and their way of life disappeared a few decades after Burke and Wills died—guns and germs did their usual nasty work. However, we do know something about the cultural evolution of food-processing techniques in a number of contemporary populations. Fijian food taboos provide one well-studied example. Joe Henrich and his colleagues (including me) have conducted a long-term study of life in several villages on Yasawa Island, Fiji. People there live in mainly a subsistence economy—about 80 percent of their calories come from yams, cassava, bananas, and a variety of other crops grown in their gardens, and from fish, turtles, and invertebrates harvested from the ocean. The remainder of their calories come from foods they purchase, mainly sugar, flour, and cooking oil. Many of the marine species in the Fijian diet contain toxins. Henrich has focused on one of these toxins called *ciguatera* (pronounced "singuatera"—the letter g is pronounced like "ng" in the word "sing"), a chemical produced by dinoflagellates that live on dead reefs.[26] Ciguatera toxin is fat soluble and accumulates in the tissues of consumers. It reaches the highest concentrations in animals at the top of the food chain, big predators like barracuda, shark, and moray eel, and in long-lived organisms like turtles. Ciguatera poisoning has severe, long-lasting symptoms in humans including joint pain, reversal of hot and cold sensation, diarrhea, and vomiting. It is a health problem throughout Oceania. Medical research indicates that ciguatera is especially harmful to fetuses and nursing infants. Interviews with people on Yasawa Island reveal that most believe that pregnant and lactating women should avoid eating the species that create the greatest risk of ciguatera poisoning, that is, large predatory fish like shark and barracuda and long-lived animals like sea turtles (fig. 1.1). Moreover, these

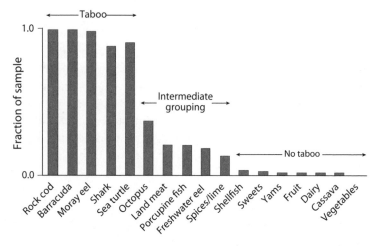

Figure 1.1. The fraction of a sample of seventy Fijian women who said that pregnant women should not eat each of the species listed. Women were presented with a list of foods and asked whether the food was suitable for pregnant women. Asking people to list foods that should be avoided by pregnant women produced similar results.

beliefs are moralized. Fijians disapprove of pregnant or lactating women who consume these species. Accordingly, I will refer to these beliefs as taboos. (The Fijians refer to them as *tabu*, pronounced "tomboo," which is closely related to the Tongan word introduced into English by Captain Cook.) Several kinds of evidence indicate that these taboos are a culturally transmitted adaptation that reduces the risk of ciguatera poisoning in the most vulnerable segment of the population.

People in the villages know that eating fish can make you sick and recognize a distinct syndrome labeled *ika gaga* (literally "fish poison"), with symptoms that are very similar to the ciguatera symptoms recognized by Western medicine. A majority of villagers suffer ika gaga at least once during their lives. They know that larger fish are riskier,[27] but they do not believe that eating turtles or moray eels can give rise to ika gaga, so the taboos are not just the result of more general beliefs about fish poisoning. They are specific to pregnancy and lactation. The taboos have an important impact because the species that are avoided during pregnancy form a significant part of the regular diet. They are fattier than other fish and marine invertebrates and provide an important source of lipids. They are also highly valued for their flavor. (I can testify that after weeks of cassava and lean reef-fish, sea turtle is *delicious*.) The taboos are also specific to those species that carry a high risk of ciguatera toxicity. Other species like parrot fish and unicorn fish are consumed much more often but are not taboo.

Most people learn these taboos from others. Henrich and colleagues asked a sample of Fijian women how they acquired their taboos. The results are shown in figure 1.2. Five percent said they came to believe that pregnant and lactating women should not consume the taboo foods based on their own experience. The rest said that someone had told them that pregnant and lactating women should not eat the

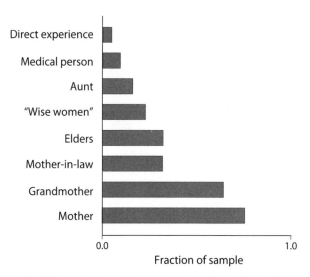

Figure 1.2. The fraction of a sample of forty-four Fijians who identified different sources of their beliefs that certain foods were not suitable for pregnant or lactating women.

taboo foods. The women who told them this were mostly family members, but about 25 percent learned from "wise women" who are explicitly recognized as experts in matters important for women—traditional medicine, childbirth, and parenting.

But why did the Fijian women believe what their parents, relatives, and the wise women told them? On the "culture-is-a-library" view, the answer should go something like this: the person they learned from gave them good reason to believe that pregnant and lactating women shouldn't eat moray eels, turtles, and the like, and they then acted on those beliefs. This belief needs to be sufficiently compelling that a woman who normally savors the rich taste of turtle will nonetheless stick to boiled cassava when she is pregnant because she is concerned about the effects on her unborn child. And it must be sufficiently compelling that she would scold a pregnant woman who is eating moray eel, even though this may offend the pregnant woman and her kin. These beliefs about the causal relationship between food and illness would be maintained because the world provides evidence that they are true. For example, somebody in the village might notice that a woman who gorged on turtle before she knew she was pregnant subsequently gave birth to a child with developmental problems.

But this account does not fit the facts. Henrich and his team interviewed many women about what they believed would happen when a pregnant or lactating woman consumed one of the taboo species. There was little consensus about the answer to this question. Many women said they did not know, and others said nothing would happen. A majority thought there would be negative health consequences for the baby, but there was little agreement about what those consequences might be—"rough skin" or "smelly joints" were among the answers that women provided. This pattern is consistent with the idea that people adopt a taboo because most people believe it and then, when asked to provide a reason, infer that there must be some kind of negative consequences. Note that this means that pregnant and lactating women are avoiding very tasty, attractive food even though they can't give any consistent answer about why they should not eat it.

Here's another example. I have studied the design of traditional houses in the same Fijian villages. These thatched houses, called *bure* (pronounced "booray"), have many important functions. They provide protection from the elements, a place to cook and socialize, and a bit of privacy for families. The houses vary to some extent in shape and size, but most share a common structure (fig. 1.3) that is widespread in Fiji. I wanted to know why their houses were built this way, and not some other way. To answer this question, I focused on the design of the structures, and how this design affects strength during the hurricanes that sweep across the island every five years or so. These storms frequently destroy houses. In the relatively mild hurricane of 2009, about 15 percent of the houses in the village collapsed, and older villagers remember a typhoon in the early 1950s in which every house but one blew down. Rebuilding a single house costs about six human-months of labor at a time when labor is also needed to replant gardens damaged by the storm. I interviewed eight men who other villagers thought were most knowledgeable about house building and asked them many questions about design. In addition to learning about house building, I wanted to know whether they could explain why the standard house design was better than alternative designs. So at the end of the first few interviews, I showed the men pictures of different houses built on other islands in the Fijian archipelago and asked what would happen to these houses when a hurricane arrived. They all agreed that the other kinds of houses would not be good to build, but they were vague about why this was the case. I thought these houses might be too different from the local houses, and so I described houses that were different in only one attribute—a different kind of wood, or some structural feature that I either added or removed from a drawing of a

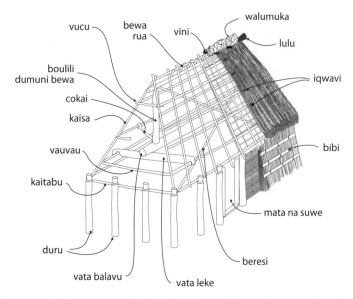

Figure 1.3. A cutaway drawing of a traditional Fijian thatched house. Labels give the name for each part in the dialect of Fijian spoken on Yasawa Island. For scale, the posts (labeled *duru*, pronounced "nduru") are about five feet high and are placed in holes that are about five feet deep (not shown). The frame is built so the thick beams inside (the *vata leke*) are above head level. The outside of the house is covered with a thick layer of thatch.

generic bure—and again asked them to compare these to the standard bure. For some features, they had clear causal explanations. One kind of wood, *noko noko*, is better for posts in dry soil, but a second kind, *vau*, is better in wetter soil near the beach. But they could give no clear reasons why many of the structural modifications I suggested were inferior to the standard design.

So it doesn't seem that people on Yasawa have good reasons to believe the food taboos, and their knowledge of house structure is incomplete. They seem to eat what they eat and build houses the way they do, in part, because this is what other people do. And I do not think that the people on Yasawa are unusual. People everywhere are influenced by the people around them. Or more formally, people seem to be intrinsically motivated to adopt the beliefs of others. Can such a psychology ever be favored by natural selection?

THE EVOLUTION OF BLIND IMITATION

In theory, the answer to this question is yes: selection can favor a psychology that causes most people to adopt beliefs *just because* others hold those beliefs.[28] I say "in theory," because this conclusion comes from a body of mathematical models that researchers like me use to study evolutionary processes. We begin with a verbal argument that captures the essential features of the process that we are interested in. In this case, we might imagine there is variation in the best behavior. For example, it might be that in some environments, pregnant women should not eat large fish, but in other environments, it's fine. In some circumstances, people might notice that women who eat a lot of large fish when they are pregnant have problems with their infants. Women could use this cue to infer that they should not eat large fish when they are pregnant. But it's also possible that it is usually difficult to detect the effects of eating large fish; after all, samples are small, and many things affect infant development. It seems plausible that a strategy that uses environmental cues when they are reliable, but depends on imitation when they are not, could be favored by selection because the occasional use of reliable cues spread by blind copying would cause the most commonly observed behavior to be the best behavior. To test the logic of this verbal argument, we build a simple model that captures its assumptions in mathematical form. Then we use mathematical tools to work out whether natural selection will favor the trait. If we end up with the same answer as the verbal argument, it's more likely that the verbal argument is cogent. If we're serious, we build a different model that captures the same basic assumptions in a different mathematical structure. If we get the same answer, then we gain confidence that the original verbal argument is right. If we don't, then we try to figure out what's wrong.

In what follows, I will walk you through one very simple model that shows that the logic of the verbal argument in the previous paragraph is correct.[29] Two reviewers of the book manuscript recommended that I cut this section because many readers find algebra intimidating. I decided to leave it in because I think it is important for readers to get a sense of how this kind of theorizing works. Do your best, but if it seems mysterious you can skip to the verbal summary at the end.

Start by assuming that there is a large population of people that live in an environment that has only two states.[30] To make things easier to follow, I am going to label these environments "large fish OK" and "large fish toxic," but keep in mind that this is not meant to be a model of fish taboos in Fiji. Its purpose is to test the logic of a general verbal argument. I assume that these people live in a variable environment because otherwise there is no need for learning. Let's assume that the environment switches states randomly in time. On average, the environment stays in a single state for a while and then switches to the alternative state. There are two possible behaviors, "eat large fish" and "large fish are taboo." Eating large fish results in higher fitness when in environments in which large fish are OK, and tabooing large fish results in higher fitness in environments in which they are toxic. The adaptive problem for the actors is to determine which environment they are in so they can choose the best behavior.

Individuals have two sources of information that can help them decide which environment they are living in. The first is an environmental cue that allows inferences about the state of the environment. You can think of this environmental cue as the correlation between the amount of large fish a mother eats and the health of her infants among a sample of people that the learner knows. To represent the verbal argument, these cues are not perfect, and they sometimes lead to the wrong inference. The cue contains information that allows learners to assess its quality. As shown in figure 1.4, the correlation is usually positive when large fish are OK and negative when large fish are toxic. So, positive correlations between infant health and the mother's consumption of large fish support the inference that large fish are OK, and negative correlations support the inference that large fish are toxic. However, the world is noisy, and as a result the observed correlation is sometimes positive when large fish are actually toxic and sometimes negative when large fish are OK. If individuals depend on a misleading cue, they will make the wrong inference and adopt the inferior behavior. Also notice that distributions in figure 1.4 are offset. The top graph tells us that an individual is unlikely to observe a large negative correlation between the amount of fish mothers eat

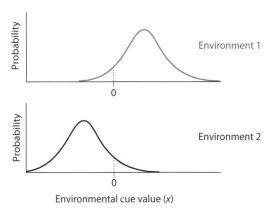

Figure 1.4. The probability of different cue values, x, in environments in which large fish are toxic and environments in which they are not. You can think of these cue values as the observed correlation between maternal diet and child health. A positive correlation is more likely when large fish are OK, and a negative correlation is more likely when they are toxic. This means that observing a positive value indicates that large fish are OK and observing a negative value indicates that they are not. However, sometimes this inference can lead to errors. The bigger the absolute value of the cue, the less likely it is that relying on the cue will lead to an error.

and the health of their offspring in an environment in which
large fish are OK to eat. Similarly, the bottom graph tells us
that an individual is unlikely to observe a large positive correlation between mothers' fish consumption and the health
of their offspring when large fish are toxic. This means that
when correlations are high (in absolute value), errors are
less likely to be made.

The second source of information comes from observing
the behavior of a sample of n other individuals in the population. These people are conventionally called "models." We
assume that on average, the better behavior is more common. So if most people eat large fish, then this is a cue that
eating large fish is OK, and if most people don't eat large fish,
this is a cue that they are toxic. But relying on this cue can
also lead to errors. It might be, for example, that most people think large fish are toxic, but by chance a learner draws
a sample in which the majority think large fish are OK. It
could also be that the environment has changed. Large fish
are now toxic, but most people think they are OK.

So individuals must decide how to behave without knowing for sure which behavior is best. They have to bet. A
branch of mathematics called Bayesian decision theory tells
us how to make risky bets to maximize the average payoff. Suppose that an individual environmental cue value, x,
is based on the correlation between mothers' diet and the
health of their offspring. The individual also observes j models in a sample of n individuals eating large fish. Then applying this theory, we maximize expected fitness by choosing to
eat large fish if:

$$j - \frac{n}{2} > -gx$$

where g is a parameter that governs how individuals weigh
the two sources of information. Although g is very important, let's ignore it for a moment and concentrate on other
parts of the inequality.

The left-hand side of this inequality will be positive when a majority of models eat large fish, a social cue that eating large fish is the best behavior. Because g is always expected to be positive, the right-hand side of the inequality will be negative if the correlation between eating large fish and infant health (that is, the value of x) is also positive, indicating that eating large fish is OK. Both sources of information point in the same direction, and expected fitness is maximized if you eat large fish. If the majority of models don't eat large fish ($j-n/2 < 0$) and the correlation between eating large fish and child health is negative ($x < 0$), the inequality tells you that eating large fish will reduce fitness, on average.

The interesting cases arise when the two sources of information conflict. For example, suppose that seven of ten models ($n = 10, j = 7$) eat large fish. This means that the left side of the inequality, $j-n/2$, is equal to 2. The observation that most people eat large fish supports the inference that large fish are OK. However, you observe a negative correlation of -0.5 between eating large fish and infant health, and this observation supports the opposite inference. In this numerical sample, the inequality says large fish are OK if $2 > 0.5g$. Thus, your decision depends on the parameter g, which governs how individuals weigh the importance of environmental and social cues. Smaller values of g mean that the social cue is more important, and larger values mean that the environmental cue is more important. In this example, if $g < 4$, then the environmental cue is ignored and the individual copies the majority of the models, but if $g > 4$, then the learner ignores the behavior of others and relies on the environmental cue.

In this simple model, the value of g is a representation of how an individual's learning psychology weights the importance of social and environmental information. Very small values of g correspond to a greater reliance on doing what others do. To determine whether natural selection can lead to the evolution of "blind imitation," which is just doing what others do, we need to let g evolve and see whether there are any conditions that favor the evolution of small values of g.

To do this, we have to model the coevolution of genes that affect learning psychology and the behaviors that provide an input to this psychology. Suppose the value of g is a genetic trait. Each individual inherits a value of g from a single parent (perhaps slightly changed by mutation but with no sex involved), acquires an environmental cue, observes the behavior of n models, and then chooses a behavior according to the rule given above. This choice determines the individual's behavior and therefore his or her fitness. People whose psychology (that is, their value of g) increases the chance of correctly identifying the environment will have more kids, and selection will increase the fraction of the population with that sort of psychology. However, the best learning psychology also depends on what people in the previous generation are doing. If most models have the correct behavior, more reliance on the social cue (smaller values of g) will be favored; if most models don't have the correct behavior, then less reliance on the social cue (larger values of g) will be favored. Genes affect what people decide, and what people decide affects the behaviors that models perform in the next generation. So the genetic evolution and cultural change are coupled, a process called "gene-culture coevolution." We iterate this system until the average of g stops changing.[31] This value tells us how natural selection will shape learning psychology in the long run.

Figure 1.5 shows the results of the model for two different sets of parameters. The raw values of g are tricky to interpret, so I have plotted the probability of acquiring the best behavior in the current environment in relation to the fraction of the population using that behavior, assuming g takes on its long-run average value. When it is easy to learn the best behavior, social information plays only a minor role in decisions about which behavior to adopt. However, when it is difficult to learn the best behavior, most individuals copy the behavior of others.

Let's think about what this means. Individuals have a learning psychology telling them the best thing to do

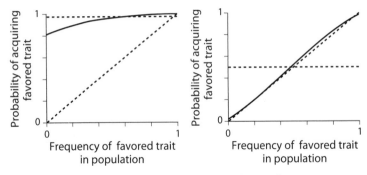

Figure 1.5. The probability of acquiring a favored trait at an evolutionary steady state when the environmental cue is high quality (left) and low quality (right) as a function of the frequency of the favored trait in the population. (Here, individuals sample the behavior of three other individuals. Larger samples favor more reliance on the social cue.) The frequency of the favored trait is the fraction of the population using the best behavior—eating large fish when they are OK and not eating them when they are toxic. The horizontal dashed line gives the probability of acquiring the favored trait using the environmental cue alone—that is, using only individual learning. In the left panel individuals have a 98 percent chance of choosing the best behavior using individual learning, while in the right panel they have a 55 percent chance of choosing the best behavior using individual learning. The dashed 45° line is the probability of acquiring the best behavior by blindly imitating a member of the previous generation. The black line gives the probability of acquiring the favored trait when using the optimal value of g at the evolutionary steady state. When individual learning is accurate, social learning has relatively little impact on fitness, but when individual learning is inaccurate, most individuals adopt the behavior they observe used by their models. In both cases environments switch every one hundred generations on average.

depending on what they have observed about the local environment. When individual learning is difficult, the environmental cue tells a little more than half of the people to do the right thing, and a little less than half to do the wrong thing. Almost all of them ignore this inference—in effect, they ignore what their own experience tells them is best—and just copy what the individuals around them are doing. Put another way, the individuals in this model are intrinsically motivated to do what others are doing. Of course, they are also motivated to follow the dictates of their experience, but this must be strong enough to overcome the motivation to imitate. Note that this is a best case for individual learning. Acquiring both cues is costless. If, as is usually supposed, individual learning is more costly than imitation, selection will favor imitation even more strongly.

A population using this decision rule has higher average fitness than a population that depends completely on individual learning, and the difference is biggest when individual learning is difficult. There are two sources of error in this model. Individual learning uses imperfect cues about the state of the environment and so is error prone. Social learning allows learners to be selective—to rely on high-quality environmental cues, and otherwise imitate. The more people imitate, the higher the average quality of the information created by individual learning. The problem with social learning is that environments change, and individuals may copy the behavior of somebody who is not well adapted to the current environment. This means that as individuals imitate more, the population tracks environmental changes less accurately. So imitation produces the biggest fitness boost when learning is very difficult and environments change slowly.

This result supports the claim that culture plays a crucial role in human adaptation. You can think of average fitness as a measure of adaptive success. The model says that culture improves adaptation most when beneficial innovations are difficult to discover, but once discovered, they are useful for a long time. It seems to me that solving difficult problems in slowly changing environments is exactly what has made people so successful. It's very hard to figure out that nardoo is toxic, and it's also very hard to figure out how to process it. The probability that any individual will be able to do both is very small. Once discovered, such knowledge is useful for a long time—archaeological evidence suggests that people have consumed nardoo in Australia for at least thirty thousand years.[32] The same goes for weaving nets and making twine, boomerangs, and fire sticks, and acquiring all of the complex lore that made Cooper's Creek a land of plenty for the Yandruwandha. In each case, the necessary knowledge is hard to acquire, and its utility degrades slowly as people move and the climate changes.

Other models tell the same story. In the model above there are only two discrete traits, but many traits vary continuously. For example, bows vary in many dimensions that affect performance: length, width, cross section, taper, and degree of recurve. We model this situation by assuming that trait values vary continuously.[33] In each environment there is an optimal design (the best combination of length, width, etc.), but this varies in different environments. In the forests of eastern North America, a long bow with a round cross section was best; but in the West, a short, flat bow was superior. Again, individuals have two sources of information: a noisy signal from the environment that tells them the optimal trait value, and the trait values of a sample of models from the previous generation. The learning problem is to combine these two sources of information.

A simple Bayesian learning model says that the best rule is the weighted average of the two signals, and the evolutionary question is how heavily each should be weighted. The answer is that selection favors placing a heavy weight on the behavior of others when the environmental signal is noisy and environments aren't changing too quickly, which means that most individuals end up with trait values quite close to the average of their models. People mainly copy, and they make substantial modifications only when there is clear evidence that they should. Historians of technology have demonstrated how such step-by-step improvements gradually diversify and improve tools and other artifacts.[34] Even "Eureka!" moments often result from lucky accidents or the recombination of elements from different technological traditions rather than from the insight of creative geniuses who think their way all the way from problem to solution.[35]

NOT-SO-BLIND IMITATION

So far, I have assumed that imitation is blind. Social learners observe the behavior of a random sample of individuals from the previous generation and then just count the number of individuals displaying each of the two behaviors. Put another way, social learners don't pay attention to the individual characteristics of the people they learn from, like their age, their gender, how successful or prestigious they are, or whether they seem honest or manipulative. I made this assumption to keep things simple, but we know from experimental work that it is very unrealistic.[36] Social learners take advantage of cues that indicate who is best to imitate. Modeling work indicates that paying attention to such cues can make social learning more efficient without requiring individuals to understand what they are doing. Consider two examples.[37]

Suppose young learners can observe the success of adults. A young girl learning to forage can see which woman brings back the most tubers, who is least tired, who is the fastest, and who other women admire. Then a propensity to imitate the successful can lead to the spread of traits correlated with foraging success even though imitators have no causal understanding of the connection. This is obvious when the scope of traits being compared is narrow. You see that your aunt is good at extracting tubers and notice that her digging stick is longer and thicker and made from ironwood. You copy all three traits, even though in reality it is just the length that makes the difference. As long as there is a reliable statistical association between length and efficiency, such nonrandom imitation will be a good idea. Causal understanding is useful because it helps exclude irrelevant traits like the decorations on the handle, but it need not be precise. Copying irrelevant traits like thickness or decoration will just add noise to the process.

Social learners must also avoid being deceived by people they learn from. Deception is not a problem for many traits. If you see a man using a bow to hunt monkeys day after day, it's likely he thinks that bows are better than, say, blowguns. However, when someone tells you that a god rewards the faithful in an afterlife, you should be suspicious and believe him only if he gives evidence that he really believes what he is telling you, for example by living a life of self-denial in service of the god. It seems likely that our social learning psychology has been shaped by natural selection so that we attend to such "credibility-enhancing displays."[38]

Note that these hypotheses about the psychology of social learning are based on evolutionary reasoning. People should be biased in favor of learning from the prestigious or those who practice what they preach, because this increased fitness in the past. Over the last decade, evidence has accumulated that supports both these hypotheses and a number of others derived using similar reasoning.[39]

COMPLEX ARTIFACTS WITHOUT CAUSAL UNDERSTANDING

All teenagers have tried to convince their parents that they ought to be able to do something because other kids do it. And if their parents are anything like mine, the kids get a maddening response: "Well, does this mean that if everybody else jumped off a bridge, you'd do it too?" It's hard to believe that our technical accomplishments can arise from such a dumb process. Don't people have to adjust their recipes in response to contingencies? The culturally evolved recipe for constructing a bow might specify Osage orange as the wood to use. But every piece of wood is different; size and density vary, there may be knots, and so on. To properly shape a bow, the bowyer has to adjust to these differences. Doesn't this require a causal understanding of bow making?

This is a common intuition, but I think it's wrong. Many animals make complex artifacts without any causal understanding of what they are doing. Bird nests, spider-webs, termite mounds, and beaver dams are just a few of the familiar constructions made by nonhuman animals, and a dip into the zoological literature reveals a long list of less familiar artifacts.[40] Many of these are highly designed and require elaborate construction techniques. Think of weaver-bird nests.

They are well designed, but observations and experiments suggest that the construction process does not depend on causal understanding. Rather, it results from an algorithm that links simple stereotyped behaviors into a sequence that generates a nest.[41] First, the bird weaves a ring, which it perches on as it constructs the egg chamber, and then it shapes the entrance (fig. 1.6). However, the evolved construction algorithm also allows individual birds to adjust their efforts in response to the availability of materials and the geometry of the branch from which the nest will hang.

Even more telling, invertebrates such as termites, funnel wasps, and spiders make complex, highly functional artifacts without any mental representation of the final form of the artifact.[42] In fact, complex artifacts can be constructed without a nervous system at all—figure 1.7 shows the "house" built by the single-celled amoeba *Difflugia corona*. The construction of artifacts is also affected by local contingencies. For example, for reasons that have to do with thermoregulation, so-called magnetic termites living in northern Australia build flat "mounds" that are oriented approximately north and south (fig. 1.8), but the exact orientation varies in response to local wind and shade conditions.[43] The millions of termites that coordinate their behavior to build these mounds in total darkness are simply executing an algorithm that causes the construction to respond sensibly to varying conditions.

Of course, executing an algorithm can go very wrong if local conditions are outside the range of variation in which the algorithm evolved. The construction of brood chambers by funnel wasps illustrates the principle.[44] These wasps brood their larvae in underground chambers dug into the side of

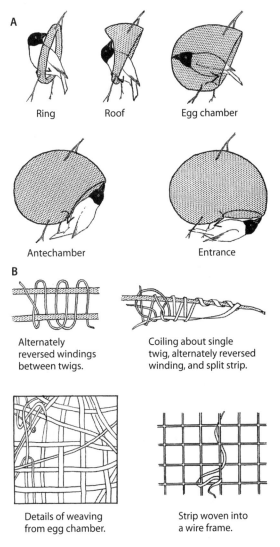

A

Ring   Roof   Egg chamber

Antechamber   Entrance

B

Alternately
reversed windings
between twigs.

Coiling about single
twig, alternately reversed
winding, and split strip.

Details of weaving
from egg chamber.

Strip woven into
a wire frame.

Figure 1.6. *A*, the construction sequence used by one species of weaverbird, the village weaver, to construct their nests. The bird first builds a hanging ring by knotting green grass stems onto the fork of a branch and then by weaving more stems to make the ring. The ring is extended outward by weaving more stems into the existing structure. *B*, a sample of the knots and weaves found in typical village weaver nests.

Figure 1.7. The "house" built by the amoeba *Difflugia corona*. It is about 0.15 mm in diameter and is made of very small grains of sand glued together. When the amoeba divides, one daughter cell inherits the house and the other constructs a new one. The genus *Difflugia* is widespread, and many of the species in the genus construct similar houses. These vary in response to environmental conditions.

a sloping bank. The entrance to the chamber consists of an upside down funnel-shaped opening that is connected to a long tube that leads to the main chamber. The tube holding the funnel has to be just the right length so that it does not touch the ground. The algorithm a wasp uses is based on construction time—after building the tube for a certain amount of time, the wasp starts the funnel. If the ground surface is raised experimentally, the wasp pays no attention even if the funnel is resting on the ground and parasitoid wasps can waltz in. Causal understanding would prevent this kind of error.

The same argument applies to things that humans make. Complex artifacts, like houses and bows, rely on causal principles that are very difficult to understand. They have many dimensions that interact in complicated ways to produce the desired goals. So copying what others do, even if

Figure 1.8. Magnetic termite mounds found in northern Australia are thin, flat structures that typically have their long axis oriented approximately north and south. However, the exact orientation depends on wind and shade conditions.

you don't understand why they are doing it, may often be a good strategy. Of course, there is some danger of a bad outcome if environmental conditions change and causal principles are violated. However, copying can be a fairly good strategy under a fairly wide range of conditions.

BUILT FOR SPEED, NOT COMFORT

Earlier I said that rodents have adapted to virtually every terrestrial habitat. They accomplished this feat through the usual combination of genetic evolution and individual learning—compared with humans, cultural adaptation plays only a small role. They choose diets, deal with toxins, avoid predators, and construct artifacts like beaver dams that rival the most complex constructions made by humans until the last few thousand years. The big difference is that there are about 2,200 species of them and only one species of us. The claim here is that we can adapt to a very wide range of environments, and other animals can't, because cultural evolution gives rise to the gradual accumulation of locally adaptive knowledge at a much faster rate than genetic evolution. A human population entering a new environment can acquire the necessary knowledge, tools, and social organization in centuries, not tens of thousands of years.

So is this claim true? Is cultural evolution really faster than genetic evolution? There is no doubt that in the modern world cultural change goes on at blinding speeds. My father grew up in a small town in upstate New York without telephones, automobiles, or electric lights. His grandchildren carry powerful computers in their pockets and fly across oceans without giving it a second thought. This stupendous rate of change is the end result of exponentially increasing rates of change that have characterized technological evolution over most of the last millennium.[45] These rates are much faster than rates of genetic adaptation in a long-lived species like humans.

A recent paper by my Arizona State University (ASU) colleague Charles Perreault provides strong evidence that cultural evolutionary rates have been much faster than genetic evolutionary rates for a long time.[46] Perreault was inspired by a famous paper in which Philip Gingrich[47] assembled paleontological data that allowed him to measure the rate of morphological evolution across time. Each point in Gingrich's data set was a pair of measurements of some morphological trait, for example femur length, for the same species at two different times. Gingrich then calculated the percent change between the two measurements and divided by the amount of time between them.

Perreault assembled a sample of 573 cases from the archaeological record (mainly a mix of foraging and horticultural populations from Holocene North America). Each point was a pair of measurements of some artifact, say the diameter of a pot, taken from the same archaeological site at two different times. Perreault then compared the measured rates of change to those in Gingrich's sample of rates of change in the paleontological record. All other things being equal, the rate of change in the dimensions of pots, points, and houses is fifty times greater than the rate of change in the dimensions of mandibles, molars, and femurs (fig. 1.9). Further, he replicated Gingrich's finding that rates of change were negatively related to the period over which

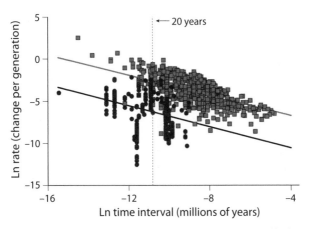

Figure 1.9. Percent change per generation for genetically heritable morphological traits (black circles) from the fossil record and culturally transmitted traits (gray squares) from the archaeological record plotted against the length of time over which the change occurred. The lines represent the best fit in a multivariate analysis of covariance. "Ln" means the natural logarithm. In both cases rates decline as the time interval increases, and the per-generation slopes are approximately equal. The distance between the lines gives the difference between cultural and biological rates of change while controlling for other variables. Cultural evolution is about fifty times faster than genetic evolution.

the measurements were made. Thus, it seems safe to conclude that culture does change much faster than genes.

WHY ONLY US?

I am often asked, if cumulative cultural adaptation is such a powerful tool, why did it evolve only once? This is a real puzzle, but one that also applies to many other major innovations in the history of life. Here are some examples. Oxygenic photosynthesis transformed the world's ecosystems by generating a rich source of energy and releasing vast amounts of oxygen. Eukaryotes, organisms with large, complex cells and a variety of other innovations, allowed the evolution of all the large, long-lived creatures that make the world interesting. Striated muscle enabled large animals to move rapidly and thus allowed for the evolution of predators and their prey. Each of these innovations evolved only once.[48]

In his superb book *Life Ascending*, Nick Lane argues that these innovations evolved only once because each was the product of chance and necessity.[49] In each case, some improbable circumstance allowed the innovation to evolve—the chance part—and then the advantages of the innovation caused it to spread—the necessity part. Oxygenic photosynthesis provides a good example. Cyanobacteria, algae, and plants combine two quite different photosystems, one a strong oxidizer to pull electrons from water to produce ATP and the second a strong reducer to push electrons onto carbon dioxide to create sugars and other complex organic molecules. Other bacteria use only one of these systems. Some use the first photosystem to generate energy from sunlight, and others use the second system to pull electrons from molecules other than water in order to make more complex organic compounds from carbon dioxide. The trick was to link these two systems together, and according to Lane, that required a rather special environment with high levels of ultraviolet radiation and lots of soluble manganese. The ancestors of cyanobacteria used manganese to protect themselves from the ultraviolet radiation, but this produced free electrons that gummed up the first photosystem. The solution to this fairly minor problem was to use the second photosystem to soak up the excess electrons. By chance, one ancestral bacterium combined both of these systems, and this nifty solution led to the evolution of cyanobacteria, algae, and plants.

So, what makes the evolution of cumulative cultural adaptation improbable? The short answer is that we don't really know. There are a number of possibilities:

Low-fidelity social learning. Lots of other organisms have enough social learning to sustain distinctive local traditions, but none show evidence of cumulative cultural evolution. For example, chimpanzees in some communities use stones to crack nuts and in others they do not, but better and better nut-cracking tools have not evolved in any community. Social cues make it more likely that certain behaviors are learned, but in every case they are behaviors that individuals can learn on their own.[50] Modeling work suggests that cumulative cultural evolution is very sensitive to the accuracy of social learning. Pete Richerson and I hypothesized that the ability to learn by observation, sometimes called true imitation, was essential for cultural accumulation.[51] It was once thought that true imitation was unique to humans. However, beautiful experiments by Andrew Whiten and his collaborators have shown that chimpanzees can learn new behaviors by observing others, although the fidelity of transmission is low compared to that in humans.[52] So it could be that high-fidelity imitation is required for cultural accumulation.

Limited potential for tool use. Other primates make limited use of tools. Members of some primate populations use stones to crack open hard-shelled nuts and mollusks, and chimpanzees also use a variety of other kinds of tools. The potential for tool use in other primates may be limited by the fact that they are all quadrupedal. This makes it hard to carry tools from one place to another, and in fact, chimpanzees make tools where they are going to be used and then abandon them. This means that costly investments in tool manufacture are less likely to pay off.

Low levels of cooperation. We will see in the next chapter that humans are much more cooperative than any other mammal. Even the smallest human societies involve webs of reciprocity and mutual aid that are perhaps fifty times larger than the interaction networks of other apes.[53] Cooperation facilitates cultural transmission for a couple of reasons. Probably most important, accurate cultural learning may require teaching. At its core, social learning is an inference problem. People don't imitate behavior. Rather, they observe behavior and then infer what people believe. Sometimes this is fairly easy. If I see you eat something, I can be pretty sure that you don't think it's poisonous. However, many kinds of social learning suffer from a "frame problem"—figuring out what part of the picture is relevant. When I see you making an arrow, a lot of things are going on. You are sitting cross legged, sharpening the arrowhead, choosing feathers to attach to the shaft, and reciting a chant. It is hard to know which parts of your behavior are essential and which are irrelevant. Psychologists like György Gergely and Gergely Csibra believe that social learning cannot be accurate unless models provide cues about which parts of their behavior exemplify generalizable knowledge and which do not.[54]

Teaching is cooperative because it provides benefits to the learner, but it is costly to the teacher. If it were practical to learn only from your parents, this would not be an evolutionary puzzle, because parents have a genetic interest in educating their offspring. But people in traditional societies learn a great deal from unrelated members of their group. Only humans exhibit much cooperation with nonrelatives.

Finally, cultural learning may rely on cooperation because cultural learning is often enhanced by spoken language. For example, a recent experiment conducted with college students suggests that verbal instruction is essential for learning how to make simple stone tools like those that appear in the archaeological record around 2.5 million years ago.[55] Students who only watched others make tools or were presented with finished tools were much less successful than those who received verbal instruction.

Language is a puzzle in itself. In most animal species, communicative signals must be costly in order to ensure their honesty. Think of the lion's roar or the red deer's antlers— these are signals that are so costly to produce that they can't be faked. A weak lion cannot make a convincing roar. But human language is "cheap talk" and is vulnerable to deception. I can say that I chased a lion from a kill when I was out hunting, even if I was cowering in the bushes. It seems likely that deception is kept under control by mutual monitoring and punishment, but this requires cooperation among individuals.[56] In contemporary human hunter-gatherer groups, the number of people speaking a language is around a thousand. No other vertebrate cooperates with nonrelatives on such scales.

Nothing useful to learn. Finally, there are good reasons to think that psychological abilities that make cumulative cultural learning possible may not be adaptive when they first arise, but only when they become widespread in the population.[57] Suppose that new bits of cognitive machinery are necessary for cumulative cultural evolution. For example, a number of authors have argued that accurate cultural learning requires the ability to create a mental representation of other people's beliefs, often called a "theory of mind." Suppose I see a snake, but a bush obscures your vision. Suppose that I don't have a good theory of mind: I won't realize that you don't know there is a snake there, so when you walk straight toward the snake I might incorrectly infer that you don't think this snake is poisonous. But if I have a well-developed theory of mind, I will realize that you have not seen the snake, so your behavior is not based on your knowledge of whether the snake is dangerous. If this reasoning is correct, a theory of mind is required for cumulative cultural evolution. Chimpanzees seem to have some knowledge of what others can see and what they know, but their theory of mind is not as well developed as ours. Let's assume that the theory of mind of early hominins was like that of chimpanzees. In this situation, rare individuals with a better-than-average theory of mind might not have been favored by selection because the population as a whole did not have the cognitive machinery necessary to accumulate locally adaptive knowledge. If this idea is correct, then theory of mind would have to evolve for some other reason like Machiavellian manipulation of other individuals. Once it evolved, cumulative cultural transmission could take off.

THE GHOST OF CULTURES PAST

The account of cultural evolution I am giving here predicts that history will influence cultural adaptation in the same way that history influences genetic adaptation. When Darwin was at Cambridge, he read and admired William Paley's book *Natural Theology*. In this book, Paley celebrates the exquisite design found in nature. In one passage, he describes in great detail how features of the eye serve the function of seeing. This, Paley thought, proved that the eye and, by a similar argument, other complex adaptations must have been created by an omniscient and omnipotent Designer. Most of us reject this argument today, but its widespread acceptance during the nineteenth century raises an interesting point. Many Victorians didn't think they had a problem explaining adaptation. A much bigger problem was the weird historical patterns of maladaptation. Why would a first-rate Designer who could create an exquisite optical device like the eye cobble up a kludge job like the human vertebral column? Why should He make the wing bones of bats more similar to the hands of moles or dugongs than they are to the wing bones of birds? Of course, now we know that evolution creates new adaptations by gradually modifying existing ones. This means that the kinds of traits favored in a lineage today depend on the history of

adaptations in the past. The wing bones of bats are modified hands, while those of birds are derived from the quite
different forelimbs of dinosaurs. The basic structure of the
vertebral column is adapted to a quadrupedal body plan. It
has been extensively modified to support human bipedalism,
but it is not how you would design a backbone if you could
start with a blank piece of paper.

History affects cultural adaptation in much the same
way. Because cultural adaptation occurs gradually over
time, variation in behavior can be partly explained by the
cultural history of populations. The behavior and beliefs of
people in societies that have been separated for only a short
time should be more similar than those of societies that have
been separated for a longer time. At the same time, different groups that have different origins may evolve different
solutions to the same environmental problem. But it is not
clear how important these effects should be. Many people
think our intelligence accounts for our ability to adapt to
a wide range of changing environments. If this is the case,
then we would predict that populations would quickly
adapt to new environments, and the ghost of cultures past
should be quickly vanquished. On the other hand, if you
think that cultural adaptations accumulate on multigenera-
tional timescales, as I do, then it will take much longer for
culture to change, and the ghost of cultures past will linger.

However, even I wasn't prepared for the results of a recent
study of the patterns of cultural variation in a large number
of indigenous societies in western North America conducted
by my Institute of Human Origins (IHO) colleagues Sarah
Mathew and Charles Perreault.[58] They based their work on
the western North American Indian data set compiled by
students of the great University of California–Berkeley anthropologist A. L. Kroeber at the beginning of the twentieth
century. Students were dispatched to reservations all around
the western United States to interview older men and women
about life before the arrival of Europeans. Kroeber had the
good sense to collect the same information from each group.

The data set contains information about 272 behavioral traits in 172 groups that lived in a very wide range of habitats covering most of western North America. The majority of these groups were small-scale societies, many based on a foraging economy. The behavioral traits recorded included technological traits (were baskets coiled?), subsistence traits (was maize grown?), traits that reflected social organization (were there matrilineal clans?), and ritual traits (were the dead buried?). After Kroeber's death, these data were curated by Harold Driver and then by his student Joseph Jorgensen, who added data on climate and the presence or absence of over one hundred plant and animal species for each group. Mathew and Perreault converted these behavioral and ecological traits into 457 dichotomous variables and added data on linguistic and geographic distances between groups. Linguistic distance is a proxy for common cultural history, and geographic distance is a proxy for cultural similarity due to the spread of cultural traits between neighboring groups.

Mathew and Perreault used these data to compare the effects of cultural history and ecology on patterns of variation in cultural traits,[59] and to examine the rate of change in cultural traits across time. In figure 1.10, the ratios of the effects of cultural history and ecology are summarized for behavioral traits grouped into a number of categories. The measure used for this comparison is calibrated so that values greater than one mean that cultural history explains more of the variation than ecology; values less than one mean that ecology is more important. You can see that cultural history is more important than ecology for all behavioral categories except subsistence, where the effect of cultural history and ecology are about the same. This means that if you want to predict whether a person makes coiled baskets, whether she belongs to a clan with matrilineal relatives, or whether her group buries their dead, it's more useful to know what language she speaks than to know what kind of environment she lives in. Amazingly, the language she speaks and where

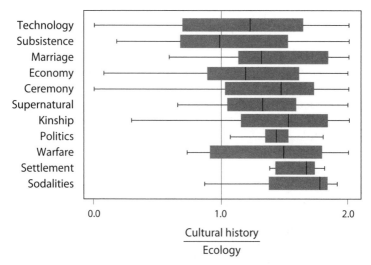

Figure 1.10. Box plots of the relative importance of cultural history and ecology in accounting for different categories of traits. For each category, the plot represents the range of relative importance value for traits within that category. Values greater than one indicate cultural history is more important; values less than one indicate ecology is more important. The vertical line in each box is the median value for traits in that category, the box is the interval between the twenty-fifth and seventy-fifth percentiles, and the error bars are 1.5 times the interquartile distance. For every category except subsistence, cultural history predicts trait values better than ecology.

she lives are equally good predictors of whether she grows maize. The ghost of cultures past haunts many aspects of people's lives.

And the ghost of cultures past lingers for a long time. To get at this, Mathew and Perreault reran the analysis using higher-level linguistic groupings as predictors. To see what I mean by higher-level groupings, think about English and its linguistic relatives. English is most closely related to Frisian languages spoken in the Netherlands and northwestern Germany. Both are descended from Saxon languages spoken around the North Sea more than a thousand years ago. Accordingly, they are grouped into a higher-level category called "Anglo-Frisian." These languages in turn are most closely related to German and Dutch, all descended from the languages spoken in an even older speech community. All four are grouped into a still higher category called "West Germanic." The West Germanic languages in turn are linked to the Scandinavian languages in a higher-level group called "Germanic." Finally, the Germanic languages are grouped with a number of other European languages like Romance and Balto-Slavic languages, Armenian, and a number of Indo-Iranian languages like Farsi and Hindi into the language phylum Indo-European, in which languages share features descended from a language spoken about seven thousand years ago, probably in Anatolia. You would expect higher-level categories to be poorer predictors of behavior than lower-level predictors because they link populations that are more distantly related culturally. This is exactly what Mathew and Perreault found (fig. 1.11). However, note that the biggest drop in predictive power is between levels 2 and 3 and between 3 and 4 in the linguistic hierarchy. These are language groupings like Salishan and Wakashan and are roughly equivalent to groupings like the Germanic and Romance families among Indo-European languages. It is thought, for example, that the Salishan languages share a common ancestor that was spoken at least a thousand years ago. This means that knowing the language spoken by a person's ancestors a thousand years ago tells you quite a bit about how she behaves today.

This work is consistent with a much larger body of work focused on "cultural phylogenetics" that has flowered over the last ten years or so. This work is aimed mainly at reconstructing the histories of particular technological traits like traditional skis,[60] and of social traits like political hierarchy,[61] marriage systems, and reproductive behavior.[62] With only a few exceptions, this work demonstrates that groups that are linguistically related are also more similar in other ways, supporting the notion that culture evolves gradually over time.

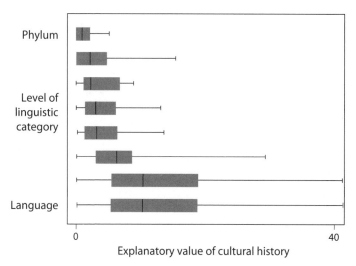

Figure 1.11. This graph plots the importance of cultural history in explaining variation in behavior using different levels of linguistic groupings. The lowest-level category is individual language, and the highest is language phylum. The biggest drop in predictive power occurs between the second and fourth levels of linguistic categorization. These are language families roughly equivalent to the Romance and Germanic groups in the Indo-European tree and suggest that cultural history acts on timescales on the order of one thousand years.

POPULATION SIZE AND CULTURAL ADAPTATION

There is another test of the idea that gradual cultural accumulation underlies human adaptation: it leads to the prediction that all else being equal, large populations will have more diverse and more complex tool kits than small, isolated populations. There are two different reasons for this prediction. First, random chance affects the number of people who adopt a variant.[63] This means that some variants will be lost by chance when their practitioners are not imitated. For instance, a wise woman's knowledge about food processing may be lost because she dies unexpectedly. The rate of loss due to this process of "cultural drift" will be higher in small populations than in larger ones because random losses are more likely. Second, social learning is subject to errors that degrade complex adaptive traits, and so most "pupils" will not attain the level of expertise of their "teachers." This creates a steady trickle of cultural loss that is counteracted by the ability of individuals to learn selectively from expert practitioners; cumulative cultural adaptation happens when a rare pupil surpasses her teachers. Joe Henrich likens this process to a runner maintaining her position on a tread-mill.[64] Learners in large populations have access to a larger pool of experts, making such improvements more likely. For both of these processes, contact between populations re-plenishes adaptive variants lost by chance, leading to higher levels of variation, and more highly adaptive traits.[65]

Michelle Kline, of Simon Fraser University, and I ex-amined the effects of population size and contact on the complexity of marine foraging tool kits among island pop-ulations in Oceania.[66] These tool kits include things like nets, fishing spears, and fishhooks. We chose island popu-lations because they are geographically bounded, making it possible to estimate population sizes and contact rates with a reasonable degree of accuracy. The groups in our sample all exploit similar marine ecosystems, minimizing the effects of ecological variation. The groups also share common cultural descent, reducing the potential impact of cultural history. We found that the total number of tool types and the complexity of individual types of tools are positively correlated with population size. High-contact societies have more tools and more complex tools than so-cieties with lower rates of contact (fig. 1.12). We tried a number of other predictors, but none explained very much of the variation among islands.

CULTURAL EVOLUTION IN THE LABORATORY

The model of cumulative cultural evolution can also be tested using laboratory cultures. I was trying to describe

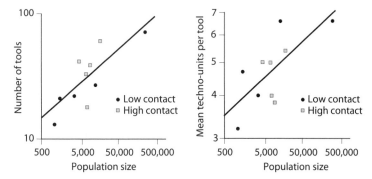

Figure 1.12. The left panel plots the number of marine foraging tools for different Pacific island societies against population size. The right panel codes the number of "techno-units" per tool as a function of population size. The number of techno-units is a rough measure of tool complexity. We used logarithmic scales because models of cultural drift and the treadmill both predict a concave relationship between population size and cultural complexity. We drew our data from ethnographic information encoded in the electronic Human Relations Area Files (eHRAF) World Cultures database. Coders estimated the total number of marine foraging tools and the average tool complexity. Low- and high-contact designation was based on coding in the eHRAF made by anthropologists who assembled the database. We included all societies from Oceania and used multiple coders who did not know the hypotheses being tested.

this work to a biologist friend of mine, and he immediately asked, "What species?" Confusion ensued, because I don't mean the kind of laboratory cultures that involve blobs on petri dishes. I am talking about experimental cultures in which information is spread socially among participants. One individual learns something and then becomes a source of information for members of a new group, and this group then becomes an information source for subsequent groups. It's the scientific equivalent of the "telephone" game that we played in elementary school. These kinds of laboratory experiments enable us to track the spread of information in small populations in a systematic way.

My IHO colleague Maxime Derex and I have used this approach to test whether social learning facilitates the cumulative evolution of complex behaviors.[67] Participants in our experiment earned money by creating totem poles like the one shown in figure 1.13c. Initially, players were provided six basic resources (fig. 1.13a) that could be combined using a workshop panel to produce higher-level innovations, which in turn could be combined to produce complex tools such as axes or rope (fig. 1.13b). These tools then allowed players to cut trees and produce logs to build their totem pole (fig. 1.13c). Other high-level innovations such as carving tools or pigments could be used to refine totem poles and increase their value. Participants' point score depended on the number of innovations they discovered and the value of the best totem pole they produced. The totem pole on the left side of figure 1.13c required 22 innovations and scored 920 points. The other totem pole required 54 innovations and scored 6,526 points.

To compare the efficacy of individual and social learning, we varied the information available to the participants. In the *individual learning* treatment, participants had access to only their own discoveries. We compared this result to that of several different social learning treatments. Here, I will focus on two of them. In the *partial information* treatment, each participant could see the innovations discovered by five other participants but could not see how these innovations were created. So, for example, one individual could see that a second individual had discovered how to make an axe but was not shown the combination of lower-level innovations that were necessary to make the axe. In the *full information* treatment, participants could see the innovations discovered by others and how to create these innovations.

The results (fig. 1.14) indicate that social information can greatly improve people's ability to solve complex problems and that the nature of the information matters. The average

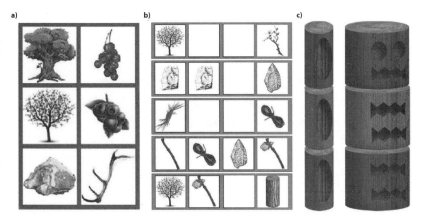

Figure 1.13. Screen capture (converted to gray scale) from the computer task used in the experiment described in the text.

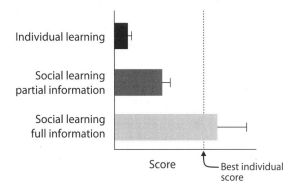

Figure 1.14. The bars give the mean point score for participants in the individual, partial information, and full information treatments. The error bars indicate one standard error, and the dashed line shows the highest score achieved by a participant in the individual learning treatment. This value is less than the score of the average player in the full information treatment.

participant in the full information treatment achieved a higher score than the very best player in the individual learning treatment. The intermediate score attained by participants in the partial information treatment shows that it helps to know that it is possible to make a particular tool, but it is not as helpful as knowing how to make the tool.

THIS CHANGES EVERYTHING

I hope I have convinced you that the human propensity to learn from others has allowed human populations to create superb cultural adaptations to local environments. Even the simplest hunter-gatherer societies depend on tools and knowledge far too complex for individuals to acquire on their own.

But some readers may be thinking, "Don't people do all kinds of things that don't seem to make much sense from an adaptationist perspective?" Every year mountaineers troop off to climb Mount Everest knowing that about one in twenty won't return. Graduate students delay childbearing to get their PhDs and then (on the remote chance they actually get a job) further delay in order to get tenure. People have beliefs about supernatural causes and agents that have no basis in fact but nonetheless lead to costly behaviors ranging from consulting psychics to eschewing sex. People identify with ethnic groups numbering in the millions, favoring insiders and discriminating against outsiders. The list goes on and on. Animals that don't bother with culture, like rats and baboons, don't seem to go in for any of this kind of thing. They behave mainly in ways that make sense. We humans, with our big brains and souped-up causal reasoning abilities, do lots of really dumb things that reduce our reproductive success.

Researchers like Leda Cosmides and John Tooby, who study psychology from an evolutionary perspective, have a stock explanation for all this wackiness that is summed up in the slogan "stone-age minds in modern skulls." They believe that human behavior is rooted in an evolved psychology that shapes choices in limited domains—how we choose mates, how we value social partners, how we think about tools, and so on.[68] This psychology contains mechanisms that respond to particular environmental cues. For example, the attractiveness of potential mates is affected by cues indicating age. These responses were adaptive in the environments in which they evolved, usually assumed to be Middle and Upper Pleistocene hunter-gatherer societies, a world of small nomadic bands with simple technology and face-to-face social relationships. Food production (herding and agriculture) led to completely novel environments that disrupted the relationship between cues and outcomes so that now our evolved psychology leads us to behave maladaptively. For example, as we will see in the next chapter, natural selection favors large-scale cooperation only among genetic relatives. Nowadays people cooperate in large groups of unrelated individuals, unlike any other mammal. This odd behavior is explained as resulting from a psychology that evolved in small-scale societies in which group members were related.

The account of the evolution of culture I have given provides another explanation for the peculiarities of human behavior. Cultural adaptation comes with a built-in trade-off. It allows people to acquire complex, hard-to-learn adaptations to local environments that they are unlikely to learn on their own. But this works only if individuals are willing to adopt the behavior of those around them even when this conflicts with their own experience and inferences. This propensity to do what others do will cause individuals to acquire *any* common behavior as long as it is not clearly contradicted by their own inferences. This means that if there are cognitive or social processes that make maladaptive ideas common, people will adopt these ideas, and as a result, maladaptive ideas can spread.[69] Let me sketch a few examples.

Evolutionary psychologists assume that natural selection shapes human psychological machinery so that individuals behave adaptively, and this means that our evolved psychology should make us "careful shoppers" in the marketplace of ideas. We should adopt beliefs that would have increased fitness in the environment in which our brains evolved. I think this is right, and it's probably the main reason that cultural evolution leads to the evolution of adaptive behavior, on average. However, there is no reason that learning mechanisms have to favor adaptive behavior in any particular case. Mechanisms that are generally useful can sometimes lead to outcomes that are different from what you might predict based on adaptive reasoning. For example, weak cognitive biases that would have no detectable effect on individual decisions can have very strong effects when repeated over "generations" by cultural transmission. This is illustrated in laboratory studies of cultural learning. University of California–Berkeley cognitive scientist Tom Griffiths and his colleagues presented experimental participants with fifty pairs of numbers and asked them to learn the rule that generated the association between the values of x and y. (For example, if I gave you the following pairs of numbers: 2 and 4; 5 and 10; and 33 and 66, you would infer that $y = 2x$). In one experiment, the x-y pairings were drawn from a parabola, as shown in figure 1.15. Griffiths and his colleagues used the learned behavior of the first subject to train a second subject, whose learned behavior was used to train a third subject, and so on, a structure that psychologists call a transmission chain. The results of one experiment are shown in figure 1.16.[70]

As you can see, the final result looks nothing like the original parabola. It turns out it doesn't matter what relationship

Figure 1.15. The dark dots give the training set of x-y pairs in one experiment by Griffiths and colleagues. After seeing the training set, one pair of points at a time, participants were given fifty values of x and asked to generate y values according to the same rule as in the training set. The light dots represent one subject's responses. This is a difficult task, but there is no obvious bias in the results.

Figure 1.16. Results of a transmission chain study by Tom Griffiths and colleagues. In the first (rightmost) trial the dark dots are pairs provided by the experimenter. In subsequent trials, the dark dots represent the training pairs generated by the previous learner. The gray dots are the pairs inferred by the current learner.

you start with; the transmission chain always ends up with a positive linear relationship because people have a weak bias to see linear relationships. (That's why you found my simple example so easy!). Weak cognitive biases may have important consequences outside the laboratory. A number of authors have suggested that supernatural beliefs arise through an analogous process. People are biased, for example, to prefer explanations based on the action of agents, and when this relatively weak preference is amplified by repeated cultural transmission, it may lead to shared beliefs in ghosts and black magic.[71]

Social learning biases that are adaptive on average can also lead to maladaptive outcomes in particular domains. We saw earlier that people's success provides indirect evidence about whether it is useful to imitate them, and both data and theory suggest that a tendency to imitate prestigious individuals is an important part of our social learning psychology.[72] However, what constitutes success varies from place to place. Here, it is courage in battle; there, it is political subtlety. This means that natural selection cannot equip people with a universal prestige psychology. People have to learn what is prestigious in the local environment. One useful cue is what the people around you find prestigious. Imitate the people that other people are imitating. This is a good general rule, but sometimes it can lead to a positive feedback process in which markers of prestige become more and more exaggerated.[73] This may explain the cultural evolution of maladaptive prestige systems in which people risk life and limb to summit icy peaks or forgo childbearing in order to publish papers in *Nature*. A bias to imitate those making credibility-enhancing displays can also have unintended side effects, like the spread of belief systems that require believers to engage in costly public behaviors like taking vows of celibacy or ostentatious poverty.

Perhaps the most surprising and counterintuitive effect of cultural adaptation is that it can cause large groups of unrelated people to engage in high-cost cooperation like warfare. This kind of behavior is not seen in other mammals, and it likely played a crucial role in making our species so successful. I will devote the second chapter to this topic.

BEYOND KITH AND KIN

Culture and the Scale of Human Cooperation

MILTON FRIEDMAN'S PENCIL

In the 1980 PBS documentary *Free to Choose*, the economist Milton Friedman used an ordinary pencil to illustrate the extraordinary level of cooperation we see in the human species:

> Look at this lead pencil. There's not a single person in the world who could make this pencil. . . . The wood from which it is made, for all I know, comes from a tree that was cut down in the state of Washington. To cut down that tree, it took a saw. To make the saw, it took steel. To make steel, it took iron ore. This black center—we call it lead but it's really graphite, compressed graphite—I'm not sure where it comes from, but I think it comes from some mines in South America. This red top up here, this eraser, a bit of rubber, probably comes from Malaya. . . . This brass ferrule? I haven't the slightest idea where it came from. Or the yellow paint! Or the paint that made the black lines. Or the glue that holds it together. Literally thousands of people co-operated to make this pencil.[1]

Friedman isn't being very scholarly here. Pencils are made from incense cedar that is virtually all from California, the lead is a mixture of graphite and clay (most graphite comes from China and India), and ferrules are typically aluminum.[2] But that is exactly his point. You don't have to know—the magic of the market takes care of all the details.

It's a matter of debate how magical the market really is, but there is no debate that specialization and exchange play a vital role in modern human societies. Virtually every aspect of your life—your clothes, shelter, food, and transportation—depends on a complex worldwide web of specialization and exchange.

Specialization and exchange are fundamentally cooperative.[3] Specialization increases the efficiency of production of goods and services, and exchange allows these benefits to be shared. However, exchange also creates opportunities for free riding. This is because there is a time lag in all but the simplest exchanges. You produce something now, and I compensate you later. This means I always have the opportunity to take your product and not compensate you. And this is exactly what happens when contracts are hard to enforce on the margins of the economy. In Los Angeles, unscrupulous firms hire undocumented laborers who work for a few weeks and are then dismissed without compensation.[4] In many exchanges there are also information asymmetries; the seller often knows more about the product than the buyer, and when this is the case exchange proceeds only if the buyer can trust the seller.

There is less opportunity for specialization and exchange in smaller societies. Nonetheless, both play a crucial role in nomadic foraging societies in which people typically live in bands numbering between twenty and forty people. In virtually every foraging group that has been carefully studied, food is shared widely in the band. In a landmark study of two South American foraging groups, Hillard Kaplan, Kim Hill, and their collaborators measured the contributions to subsistence made by men and women of different ages.[5] The results are shown in figure 2.1. Young

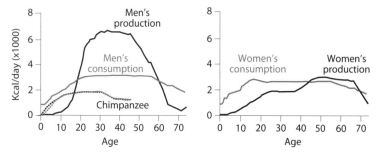

Figure 2.1. Male and female production and consumption of food resources measured in kilocalories/day plotted against age in two contemporary foraging groups, the Hiwi and the Aché, and similar data for chimpanzees. Male and female chimpanzees produce virtually all the calories they consume after weaning, while in these human populations middle-aged males produce food that is transferred to members of other age and sex classes who engage in other productive activities.

men and older men consume more than they produce, while middle-aged men produce large surpluses. Young women also consume more than they produce, while middle-aged and older women generally break even. The foraging economy of these groups is rooted in specialization and exchange. Middle-aged men specialize in food production and provide a net flow of calories to people in other age and sex classes, especially reproductive-aged women. Middle-aged women specialize in child care, food processing, craft production, and other essential tasks. It seems likely that this kind of food sharing and exchange has gone on long enough that it has shaped the evolution of human life history. Humans are unusual in coupling slow development rates with relatively short interbirth intervals. This means that they may be caring for multiple dependent children at the same time. Other ape mothers do not do this. Kaplan, Hill, and colleagues provide convincing evidence that specialization and exchange make this life history profile possible.

And it's not just food sharing. People in foraging societies exchange a wide range of services. According to Kim Hill,

> non-foraging cooperation in Aché forest camps includes services such as clearing a camp spot for others; bringing water for others; collecting firewood for others; lighting or tending another's fire; cooking and food processing for others; building a hut that others share; making, fixing, and lending every imaginable tool; grooming others; keeping insect pests away from others; tending others' infant and juvenile offspring; feeding another's offspring; teaching another's offspring; caring for others when they are ill; collecting medicinal plants for others; listening to others' problems and giving advice; providing company for others who must stay behind in camp or go out to forage alone; and even entertaining others (singing, joking, telling stories) when requested.[6]

Hunter-gatherers gain from cooperation in myriad contexts.[7] Very few mammals show any evidence of specialization or exchange despite the big gains that they make possible. Figure 2.1 also plots the calorie production and consumption for chimpanzees—after weaning, each chimpanzee produces about the same number of calories it consumes. There is virtually no exchange—each adult chimpanzee collects and processes almost all of its own food and makes its own nest each night. Although chimpanzees sometimes share meat, this constitutes a small fraction of the diet.[8] This is typical for mammals. Cooperative breeding has evolved a few times in mammals—for instance in naked mole rats, banded mongooses, meerkats, tamarins and marmosets, wild dogs, wolves, and beavers—but this form of life is uncommon.[9] Moreover, in many cooperatively breeding mammals there is likely little increase in efficiency. Dominants suppress the reproduction of subordinates, who are left with little choice but to help raise the dominants' offspring, and there is little or no gain in average fitness.

Outside the mammals, many species have lots of specialization and exchange. But before I discuss them, let's deal with another form of cooperation common in human societies but absent among most other mammals: the production of public goods.

YOU DIDN'T BUILD THAT

In 2012 in Roanoke, Virginia, President Obama gave the speech with his infamous (at least to the viewers of *Fox News*) "You didn't build that" remark. It seems likely that Obama was inspired by Elizabeth Warren's riff in a viral YouTube video.[10] Here's what Warren said:

> There is nobody in this country who got rich on their own. Nobody. You built a factory out there—good for you. But I want to be clear. You moved your goods to market on roads the rest of us paid for. You hired workers the rest of us paid to educate. You were safe in your factory because of police forces and fire forces that the rest of us paid for. You didn't have to worry that marauding bands would come and seize everything at your factory.

Warren is pointing to another crucial form of cooperation, the production of public goods. Things like roads, public education, public order, and collective defense create very large benefits that are shared by all members of a modern society whether or not they contribute to the costs of producing these goods. As we all know, this motivates people to free ride, to take the benefits of public goods without contributing to their production. For some goods like roads, education, and collective defense, modern states have institutions such as taxation and the draft, backed by criminal penalties, designed to motivate people to contribute, but for other goods like public radio, people contribute voluntarily. The existence of all modern societies depends on the production of public goods.

Public goods also play an important role in small-scale hunter-gatherer societies. Foragers make shared investments in improving the local environment through burning and other habitat modifications; they construct shared capital facilities like drive lines and fish weirs; they participate in warfare. The provision of these public goods often involves the cooperation of hundreds of individuals, so each person makes only a small individual contribution.

The production of public goods plays a central role in every human society but is rarely observed in other species. Chimpanzees do not build fences to keep out their neighbors, and baboons do not form babysitting cooperatives, even though these efforts might be beneficial for everyone. In colonies of bees and ants, when public goods were produced, large-scale group-level cooperation evolved because cooperators were genetic relatives. Human cooperation is different. In even the simplest human societies, people cooperate in large groups of almost unrelated individuals to provide public goods. In this chapter, I will argue that this is possible because behavior in human societies is regulated by culturally transmitted social norms. Within a culture, people share beliefs about what is right and wrong, and these beliefs motivate systems of punishment and reward that influence behavior. We saw in the last chapter that cultural adaptation means that people have to be motivated to acquire the beliefs of people around them. In this chapter, we will see how this gives rise to norms that allow people to produce public goods. At the end of the chapter, I will argue that culturally transmitted moral beliefs also explain why specialization and exchange are so much more important for humans than for other animals.

LARGE-SCALE PUBLIC GOODS IN SMALL-SCALE SOCIETIES

I am going to spend some time here documenting large-scale public goods production in hunter-gatherers because many authors believe that large-scale cooperation is a recent phenomenon and that in Pleistocene[11] foraging societies, cooperation was limited to small groups and motivated by kinship and reciprocity alone. The fact that large-scale cooperation has been widely observed in foragers counts against this view and supports the alternative argument that cultural adaptation has made people supercooperators for a long time.

People in many foraging societies undertake activities aimed at increasing the productivity of the local habitat. The use of fire to create more productive plant communities is a very widespread example. The plant communities that emerge after burning are dominated by fast-growing species that yield higher animal biomass.[12] For example, the Mardu, an Aboriginal group living in Australia's Western Desert, set fires in grasslands during the winter season.

Rebecca Bliege-Bird and her collaborators have shown that this creates habitats with higher foraging returns for small game like monitor lizards.[13] As these authors point out, the environmental changes induced by burning are likely to be public goods. The individuals who manage the burning experience costs, and everybody benefits. But burning is a low-cost activity, so it is possible that the individuals who manage the burning gain enough benefits themselves to off-set their costs. If so, then this is not a good example of a public good. However, foragers also make many more costly investments in habitat improvement. For example, Native American groups along the Mississippi and the Colorado Rivers sowed the seeds of wild grasses on mudflats exposed after seasonal floods.[14] Other groups transplanted tubers and fruit trees. The Aché of Paraguay cut down trees and returned months later to harvest beetle larvae from the dead tree trunks.[15] Perhaps the most spectacular example comes from the Owens Valley Paiute in California, studied by Julian Steward (fig. 2.2).[16] The Paiute built diversion dams and canals to irrigate land and increase the growth of water-loving plants with edible roots. The largest of these irrigation areas covered about ten square kilometers and was fed by canals that were several kilometers long. Construction required the cooperative effort of virtually the entire local population.

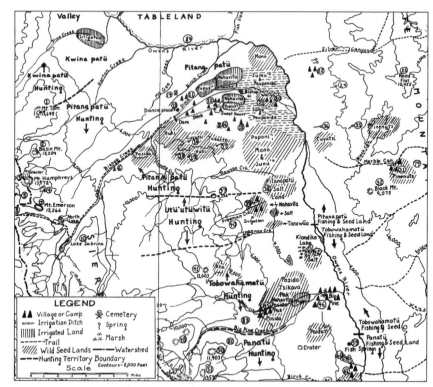

Figure 2.2. A map of artificial wetlands created by the Owens Valley Paiute during the nineteenth century. The large central area on Bishop Creek involved two canals several kilometers long and is very close to the modern city of Bishop, California. The smaller areas on Pine Creek and Baker Creek were irrigated using shorter canals. The Paiute harvested bulbs from a number of plant species that grew in these artificial wetlands. Such bulbs often required extensive processing.

Foragers also invested in large-scale building projects to aid in the capture of game. Drive lines are fence-like constructions that concentrate animals like caribou, buffalo, bighorn sheep, and pronghorn so that they are more easily killed. These were widely used in the Great Basin for at least three thousand years. The Whisky Flat pronghorn trap in Nevada north and east of Mono Lake provides a

well-studied example.[17] A fence 2.3 kilometers long channeled the pronghorn into a large circular corral where they
were easy prey for hunters armed with bows. The fence
and corral were built from about five thousand juniper
posts. It must have taken many, many people to fell all
the trees, dig all the post holes, and collect and place all
the bracing stones. Similar drive lines have been found
throughout the Great Basin. Plains Indians built mul-
tikilometer drive lines that funneled stampeding buffalo
herds over cliffs.[18] The Inuit used drive lines to hunt caribou throughout the Canadian Arctic. These were sometimes several kilometers long and were constructed from
cairns of stones called *inukshuk* by the Inuit.[19] The Dorset,
people who lived in the same area before 1000 CE, built
shorter, funnel-shaped drives, probably because they did
not have bows and needed to get closer to kill the caribou
with lances (fig. 2.3). Similar stone drive lines have been
found in Alberta and Michigan, the latter dating to about
nine thousand years ago.[20]

Figure 2.3. Drive lines from the POD site on Victoria Island,
Nunavut. Two drive lines create a funnel that channeled autumn
caribou herds into a narrow space between two pits that protected
Dorset hunters armed with lances. The sections adjacent to the
pits were solid walls (between 17 and 40 meters in length). Beyond
that the drive line was made up of closely spaced cairns.

Figure 2.4. A salmon weir built across Mill Creek in northern California by the Hupa.

Coastal and riparian foragers all over the world constructed weirs (barriers in rivers and streams) to harvest fish. These varied in size from small fish traps to large constructions that spanned substantial rivers. Figure 2.4 shows a weir across Mill Creek in northern California built by Hupa people in the first part of the twentieth century. We don't know much about its construction, but we know a lot about a similar weir built by the Yurok across the nearby Klamath River at around the same time.[21] Stout wooden pilings were driven into the riverbed and then fencing was added to block the upstream movement of the yearly salmon run. Men with dip nets harvested the fish when they were forced through small openings in the weir. More than 150 men were needed to cut the timber, and about 70 to build the weir. Amazingly, the entire structure was dismantled after only ten days of use, probably to allow salmon to continue their run and maintain peace with their upstream neighbors.

Intergroup conflict also creates important public goods. Modern battles like the Battle of Stalingrad can tip the balance between defeat and victory, with defeat sometimes leading to widespread pillage, rape, and subjection. Many hundreds of thousands of men risked their lives in such battles, so each individual had only a minuscule effect on the outcome. Hero and coward alike experience the sorrows of defeat or the fruits of victory, and this means that all other things being equal, self-interested individuals should do what they can to avoid battle. Of course, in modern battles all other things are not equal, and modern nations have evolved complex institutions to motivate soldiers to fight.

There is debate about whether intergroup conflict also creates important public goods in foraging societies. The key problem is the size of war parties. If they are small enough, then each individual can have a substantial effect on the outcome. For example, chimpanzee border patrols typically number from five to fifteen individuals, and chimpanzees never launch attacks unless they greatly outnumber their opponents. If the costs of conflict for each individual are very low, then each participant's share of the benefits associated with increased territory size could exceed the costs of participation. There is no doubt that hunter-gatherers engage in intergroup conflict, and that the resulting mortality rates are often high.[22] However, some authors believe that forager conflicts, like those of chimpanzees, are limited to ambushes and raids in which war parties are small, and conflict is avoided unless one side has substantial numerical superiority.[23] Others argue that foragers sometimes engage in large-scale battles with many warriors on each side and substantial numbers of casualties.[24] Modern ethnographic data cannot resolve this dispute because twentieth-century foraging groups like the Aché and the Hadza are surrounded by much more powerful farmers or herders, making warfare unprofitable. I believe that historical accounts strongly suggest that foragers sometimes engaged in large-scale conflict in which fighters experienced substantial costs. The best data come from Australia, a continent of foragers until the arrival of Europeans at the beginning of the nineteenth century. William Buckley, a young man transported to Australia in 1803 as punishment for receiving a bolt of stolen cloth, escaped and lived with the Wallaranga, an Aboriginal group, for thirty-five years. His account of his life among the Wallaranga is saturated with interpersonal violence on all scales, including murder, small-scale raids, and large battles in which whole tribes were mobilized.[25] War parties could number as many as three hundred fighters, and battles produced many casualties. A review of more scholarly accounts of Aboriginal life confirms Buckley's picture—war was common, war parties were sometimes large, and death rates were substantial throughout Aboriginal Australia.[26] Historical accounts of peoples in western North America tell a similar story. Linguistic evidence suggests that speakers of Numic languages like Paiute and Shoshone spread across the Great Basin during the last millennium, and at contact, groups on the periphery engaged in large-scale infantry combat with their Great Plains neighbors.[27] Later, one of these groups, the Comanche, acquired horses and mounted large-scale cavalry raids on their neighbors. The Comanche eventually achieved hegemony over a territory ranging from central Colorado to the Rio Grande.[28] Archaeological evidence suggests that warfare was also common on the Great Plains before the arrival of white settlers.[29] In the western Arctic, people belonging to different Inupiaq ethnolinguistic groups conducted regular large-scale warfare against members of other groups.[30] Warfare was common among California Indians, especially in the southeastern desert regions and among coastal people.[31] Again, small-scale raiding was common, but there are also many reports of larger conflicts involving more than one hundred warriors on each side. Among sedentary hunter-gatherers, like the sago foragers of lowland New Guinea and the salmon fishers of the northwest coast of North America, conflicts involving hundreds of individuals were relatively common. For example, A. L. Kroeber describes a battle between the Yurok and the Hupa in northwestern California in which about one hundred Yurok warriors launched a surprise attack on a Hupa settlement. About a year later, a similarly sized group of Hupa attacked Yurok settlements in retaliation.[32]

I think that historical accounts warrant three conclusions about warfare among hunter- gatherers:

1. Foraging societies were often very violent. Most intergroup conflict took the form of small-scale ambushes and raiding. However, battles involving hundreds of men on each side occurred in many foraging societies.

2. The scale of warfare was likely limited by logistics, not the ability to solve collective-action problems in large groups. For example, large-scale conflict was common in Inuit societies west of the Mackenzie River delta but rare among the Central Inuit farther east, even though these people were all Inuit speakers and shared many cultural institutions. Western populations fought in larger groups because they lived at higher population densities.[33]

3. Much of the large-scale conflict probably occurred between members of different ethnolinguistic units. If nomadic foraging bands were the size of modern foraging bands, then large war parties must have been composed of warriors from many bands. Most reports indicate that these were bands from the same "tribe," and their opponents were members of other "tribes." However, we should be cautious about this conclusion because European observers may have been biased by preconceptions based on nation-state warfare.

THE PUZZLE OF HUMAN COOPERATION

We have seen that humans cooperate in ways not seen in most other vertebrate species. Every modern human society depends on division of labor, exchange, and the production of public goods. This includes nomadic foraging societies that are likely similar to human societies over the last few hundred thousand years. Only a few other vertebrate species do any of these things. However, many invertebrates, especially social insects like ants, bees, and termites, as well as some social spiders and many microorganisms, also have specialization and exchange and produce public goods. In many ant species, a small number of individuals are reproductive specialists, while the vast majority are workers or warriors. They build complex shared nests, and some species even cultivate fungus gardens. Large-scale cooperation is also seen in microorganisms like slime molds. In fact, you can think of multicellular organisms as highly cooperative societies of cells. Cells in different tissues perform different tasks—food gathering, defense, and reproduction.

Many of the lineages that have evolved division of labor and trade have been spectacular ecological successes and have expanded into a vast range of habitats. Multicellular organisms arose when several different lineages of single-celled creatures evolved specialization and exchange, and their descendants have come to occupy a very large number of niches. Similarly, eusocial insects occupy an enormous range of foraging niches ranging from carnivory to farming, again suggesting that cooperation provides benefits in a wide range of environments. Furthermore, the eusocial insect colonies cooperate in multiple domains. Army ants, for instance, breed cooperatively, work together to build bridges, defend the colony, manage traffic, and have several castes of workers specialized for different tasks.[34]

It has been argued that we are so cooperative because we are so smart. But available evidence suggests that big brains aren't necessary for cooperation.[35] Wolves and African wild dogs engage in cooperative hunting and food sharing, but they have smaller brains than humans and other apes. The other mammals on the list of cooperatively breeding species have even smaller brains. There is a positive correlation between social group size and measures of brain size, but many species that live in large groups are not very cooperative. Moreover, among corvids, a family that includes crows, jays, magpies, and ravens, there is no relationship between brain size and measured levels of cooperation.[36] Social insects and other invertebrates achieve intricately choreographed cooperation in immense groups with extremely limited cognitive abilities. They are able to coordinate for mutual benefit not because they are smart, but because they have solved the problem of free riding.

They can do this because cooperation occurs among relatives. Most evolutionary biologists believe that natural selection can be thought of as maximizing "inclusive fitness"—a quantity that weights the effect of a change in a focal individual's behavior on the fitness of a second individual by the degree of relatedness between the focal and the second individual. Relatedness, r, gives the probability that two individuals share a gene by common descent. It is 1/2 for full siblings and parents and offspring, 1/4 for grandparents and grandchildren, 1/8 for cousins, and so forth. A behavior that decreases the fitness of an actor by an amount c and increases the fitness of another individual by an amount b will be favored when the change in inclusive fitness, $rb - c$, is greater than zero.[37] This means that when relatives interact, selection acts as if actors place a positive value on the fitness of others.[38] Most of the social groups described above are composed of close kin and so the evolution of cooperation is easy to understand. There are species of social insects in which relatedness is now low, but there is good evidence that when extensive cooperation originally evolved in these lineages, social groups *were* composed of closely related individuals.[39] Once reproductive specialization evolved so that workers could not easily reproduce, the cost of cooperation to nonreproductive individuals became very low. So the bottom line is that specialization and production of public goods are widespread in nature, and they evolved because interacting individuals were closely related.

Humans are a big exception to this rule because they cooperate with large numbers of unrelated people. Think of the huge number of people who cooperated to provide Milton Friedman with his pencil—lumberjacks in northern California, miners in China, engineers in Washington running the dams that generated the electricity used to smelt the aluminum in the ferrule, and so on and on. The same goes for public goods production. The million Russians who labored in the snows of Stalingrad were virtually unrelated. Contemporary human societies are testimony to the human ability to cooperate with very large numbers of unrelated people.

The same goes for human hunter-gatherers. People often picture hunter-gatherers cooperating in small bands made up mainly of closely related individuals. However, two recent papers by my colleague Kim Hill and collaborators make it clear that this picture is wrong for ethnographically well-studied foragers. First, band members are not particularly closely related. A survey of thirty-two hunter-gatherer societies[40] indicated that on average primary kin make up less than 10 percent of band members. In one especially well-studied population, the Aché, the average relatedness in sixty bands was 0.054, which means that benefits have to be about twenty times the cost for altruistic behaviors to be favored by natural selection. Second, mutually beneficial social interactions are not limited to band members but extend to virtually all members of an ethnolinguistic group.[41] We have already seen that hunter-gatherers can mobilize people from many different bands to construct drive lines, fish weirs, and irrigation works, and they can muster war parties of hundreds of fighters. Detailed quantitative data on interband cooperation among two contemporary foraging groups, the Aché and the Hadza, support this claim. The northern Aché number about five hundred individuals, and the Hadza more than one thousand. Hill and his colleagues conducted interviews with the Aché about their interactions with other members of their ethnolinguistic group. He showed people pictures of other individuals and asked them how long it had been since they had engaged in various types of cooperative behaviors with them, such as sharing food and child care. Yale anthropologist Brian Wood conducted a similar set of interviews among the Hadza.[42] He then used these data to estimate rates of interaction for each type of behavior. The results suggest that the Aché and Hadza interact cooperatively with most members of their ethnolinguistic group about once every two years on average, and by extrapolation they interact with them many times over the course of their lives. Of course, they cooperate much more often with fellow band members than with people from other bands, but band membership is extremely fluid, and people frequently move from one band to another. In most other primate species, including all ape species, peaceful social interaction takes place within residential groups, and interactions between members of different groups are typically hostile. Human foragers are different. They live in residential groups similar to those of other primates, but unlike most other primates, people have peaceful, often cooperative, interactions with a much larger group that shares their language and customs.

The average degree of genetic relatedness within hunter-gatherer ethnolinguistic groups is usually quite low. A standard formula from population genetics says that when groups are large, when migration rates are low, and when migrants travel a long distance before settling, the average relatedness within groups, r, is approximately given by

$$r \approx \frac{1}{1 + 4Nm}$$

where N is the group size and m is the migration rate per generation.[43] With this formula, it's easy to see why average relatedness in most primate social groups, including humans, is low. Social interactions take place within residential groups that number around one hundred for ground-dwelling species like chimpanzees and baboons.

The members of one sex emigrate at sexual maturity, and there are roughly two generations present in the group, so migration rates are about 25 percent per generation. This means that $r \approx 0.01$. What about human hunter-gatherers? Social interaction takes place at the level of the ethnolinguistic group, which numbers around five hundred and has intergroup migration rates of around 0.05, which means that $r \approx 0.01$. We should expect somewhat larger values because of corrections for local migration, overlapping generations, variation in group size, and the effect of two sexes, and two recent genetic studies report values of $r \approx 0.03$[44] and $r \approx 0.07$.[45] So, theory and data suggest that the value of r for both chimpanzees and humans is less than 0.05, and this means that contributions to a public good will be favored by selection only if the benefit to an individual is more than twenty times the cost of that individual's contribution.

Three or four million years ago, our ancestors lived in apelike societies. Groups were small, relations between members of different groups were hostile, and there was little exchange and no production of public goods. By the time that modern human behavior emerged, probably between one hundred thousand and two hundred thousand years ago, people were committed to cooperative exchange and the production of public goods in ethnolinguistic groups of roughly one thousand mostly unrelated people. Everywhere else in nature, large-scale cooperation is explained by kinship, but in humans it is not. So here's the puzzle: How could natural selection favor changes in human psychology that led to cooperation among large numbers of unrelated people?

RECIPROCAL ALTRUISM IS NOT THE SOLUTION

A few years ago I shared the stage at a public panel discussion with a young evolutionary psychologist. I spent my allotted minutes describing how people cooperate in large, unrelated groups, posing it as an evolutionary puzzle. When his turn came, this fellow said something like, "There are

two explanations for the evolution of cooperation: kin selection and reciprocal altruism. So, reciprocal altruism must
explain human cooperation." While they rarely put it so
baldly, this view is shared by many who seek to understand
human behavior within an evolutionary framework.

There is a sense in which this fellow has to be right—but
only if you take "reciprocal altruism" to mean any kind of
behavior in which individuals make current choices contingent on the past behavior of their social partners. However,
many people interested in human evolution take reciprocal
altruism to mean contingent cooperation—I'll cooperate
now if you cooperated in the past. We will see that reciprocal altruism, defined this way, can sustain cooperation
only in small groups. Cooperation in larger groups requires
systems of norms enforced by sanctions imposed by third
parties, and we will see that there is a very big difference
between such norms and reciprocal altruism. It is important to understand why reciprocal altruism, in the contingent
cooperation sense, cannot explain the evolution of observed
patterns of human cooperation.

The evolutionary theory of reciprocal altruism began
with a nonmathematical paper by Robert Trivers.[46] The
basic idea is simple: I help you in expectation that you will
help me in the future. If you don't help me when the time
comes, then I will stop helping you. As long as the benefits
of a long-term partnership exceed the short-term gain associated with cheating, selection will favor cooperative behavior. This can work only if individuals interact repeatedly,
can recognize their partners, and remember their previous
interactions. Clearly reciprocal altruism is not in the least bit
altruistic—if cooperators are successful, it is because they
get paid back in the long run. So a much better label is "reciprocity," and that is the term I will use here. At the end of his
paper, Trivers mentioned that reciprocity could be modeled
using the iterated prisoner's dilemma.

Ten years later, this suggestion was taken up by Robert
Axelrod and W. D. Hamilton.[47] They took an evolutionary

game theory approach. Pairs of individuals are sampled from a population and interact repeatedly. During each time period, individuals can cooperate or defect. In the version of the prisoner's dilemma favored by biologists, a player who cooperates produces a fitness benefit b for the other individual at a fitness cost $-c$, and individuals who defect produce no benefit and experience no cost. In a one-shot interaction, you are better off defecting no matter what your partner does (if she cooperates, $b > b-c$, or if she doesn't, $0 > -c$). When interactions are repeated, contingent behavior can sustain cooperation. For example, suppose individuals follow a rule, called a "strategy" in the jargon of game theory, named Tit-for-Tat: cooperate on the first interaction and then do whatever the other player did in the previous interaction. When two Tit-for-Tat players are paired, they cooperate until they stop interacting. When a Tit-for-Tat player is paired with a player who always defects, the Tit-for-Tat player stops cooperating after the first interaction. If Tit-for-Tat players are common in the population, a Tit-for-Tat player will interact mainly with other Tit-for-Tat players and will receive the long-run benefits of sustained cooperation. Defectors will receive a bigger benefit, but only for a short while because their partners will no longer cooperate with them. As long as interactions go on long enough that the long-term benefit of mutual cooperation exceeds the short-term benefit of exploiting a cooperator, Tit-for-Tat has higher fitness than defecting.

Axelrod and Hamilton's paper led to an avalanche of research on the iterated prisoner's dilemma. This literature teaches four main lessons about the evolution of reciprocity:[48]

1. When unrelated individuals interact and interactions go on long enough, populations in which contingent cooperation is common can resist invasion by defecting strategies. We say such strategies are "evolutionarily stable strategies" (in the jargon of evolutionary game theory, an ESS). And "long enough" is not

very long. For example, if $b/c = 2$, individuals have to interact only twice on average. There are a vast number of cooperative strategies, and most are evolutionarily stable if interactions persist.

2. When unrelated individuals interact, noncooperative behavior is also evolutionarily stable as long as the behavior is costly $(c > 0)$.

3. If interacting individuals are related, contingent cooperation can invade a population in which unconditional defecting is common. And the individuals don't have to be very closely related. For example, if $b/c = 2$ and individuals interact eight times, contingent cooperation can invade unconditional defection as long as $r > 1/16$, which is about the average level of relatedness in Aché bands.

4. It is much more difficult for one cooperative strategy to invade a second, less efficient cooperative strategy than it is for it to invade a strategy that always defects.[49] This means that relatedness that allows crude reciprocal strategies to get started in a world without cooperation may not allow further evolution of more effective reciprocal strategies.

These mathematical results support Trivers's verbal reasoning. This theory predicts that reciprocity will be common in social mammals and birds that interact repeatedly, recognize group members as individuals, and are cognitively sophisticated.

SIZE MATTERS

There are two big problems with reciprocity as an explanation of human cooperation. First, the theory seems to predict that reciprocity will be very common among cognitively sophisticated social animals, but nature stubbornly refuses to agree. Although there is some debate, most behavioral ecologists think there are very few good empirical examples of reciprocity among other animals.[50] Primates may be an exception, but even there the evidence is not overwhelming. This suggests there is something fundamentally wrong with the theory. We will return to this problem later. Here I want to focus on a second, much bigger problem when it comes to explaining human cooperation—contingent cooperation can sustain the provision of public goods only in small groups.

To see why, imagine that you are a forager who lives in the Australian desert and shares a number of water holes with several hundred other people. Every once in a while people from neighboring groups attempt to seize one of the water holes you use. To prevent this from happening, people in your group have to join together and drive off the invaders. Sometimes the invaders resist, and this can lead to lethal combat. Everybody is afraid, but having the water hole is worth the risk if everybody fights. Now suppose one person in your group runs away when the fighting starts, but the rest of you manage to drive off the invaders. The coward reduced his chance of getting killed but still gets to use the water hole. Defense is a public good.

Now suppose people in the group are contingent cooperators who use the rule "I'll fight as long as everybody else fights." A single coward causes everyone else to avoid fighting too. When the coward runs, he is worse off because nobody defends the water hole. Knowing this, he will decide to fight. In theory, this can allow cooperation to be evolutionarily stable. In reality, two big problems arise. The first problem occurs if there is uncertainty or error about who is cooperating. A brave warrior might mistakenly think that a signal to retreat was given and turn back in error, or somebody might mistakenly think that a courageous fighter ran away, when he was actually hidden from their view. It seems likely that these kinds of errors will happen often, and each time they will produce a wave of defection. A second problem is that strategies of this kind can never be favored by selection when they are rare. If the world is composed mainly of defectors, then cooperation cannot get started unless groups are very small because there is essentially no chance that any group will have enough cooperators to sustain cooperation even if groups are composed of close relatives.[51] Strategies that tolerate a few defectors help a little but don't really solve the problem.[52]

Direct sanctions work much better. Suppose that instead of refusing to cooperate, people impose sanctions on cowards directly. Such sanctions could take many forms, some costly, some not. You and the other cooperators could physically punish the cowards. The Tiwi of Australia punish offenders by throwing spears at their thighs.[53] This is costly because it takes time, and offenders may fight back. You could ostracize the offenders by denying them access to water holes and other benefits of living in the group's territory, a form of punishment used by the Hadza.[54] This may be costly if offenders attempt to use group resources anyway and you are forced to take action against them. Finally, offenders can be denied the privileges that are sustained by various forms of reciprocity. The next time cowards are ill or injured, nobody feeds them; when they come home empty-handed from a hunt, no one gives them meat. Sanctions that involve withholding help are actually beneficial to prospective helpers in the short run because they allow them to avoid the normal costs of social obligations without loss of reputation.[55]

There is a terminological confusion here that needs to be dealt with. Some authors, especially in evolutionary biology, use "punishment" only when sanctioning is costly in the short run and prefer to use the label "negative quasi-reciprocity" for indirect sanctions that involve withholding benefits.[56] Others use "punishment" for any sanctions that can be targeted at a particular defector[57] and then specify whether the punishment is costly or not. I prefer the latter usage because it conforms to the everyday meaning of the word. For example, we punish children by taking away privileges to which they are normally entitled. This can sometimes benefit the punisher—you don't have to drive to the mall when your teenager is grounded. However, to prevent confusion I will use the term "punishment" only when the sanctions are costly to impose.

There are several reasons that direct sanctions are more effective than reciprocity in public goods settings. First, they can be targeted at defectors. Errors or uncertainty do not set off cascades of defection, and this means that cooperation can be stable in a noisy world. Second, a minority of punishers can guarantee that cooperation pays. If the cost of being sanctioned is larger than the cost of cooperating, even a modest number of sanctioners can motivate others to cooperate. This means that sanctioners may be able to motivate others to cooperate even when the sanctioners are fairly rare in the population.[58] Third, sanctions act as a deterrent. If the threat of sanctions motivates people to follow the rules, then sanctions will be imposed only when people defect in error, and this means it may not cost much to prevent defections. And carrots are more expensive than sticks. If rewards are used to induce people to cooperate, they will have to be paid every time cooperation is performed. Finally, with reciprocity the maximum cost to the individual being sanctioned by a single actor is limited to the loss of benefit associated with the withdrawal of cooperation by that actor. The magnitude of direct sanctions can be as much damage as a sanctioner can afford to administer.

All of this argument fits with what we all know about cooperation in everyday life. Cooperation on small scales is regulated by reciprocity. If you invite me to your house for dinner and I don't reciprocate, I won't be invited again. If you belong to a small committee and some of the other members stop contributing, you may do the same. Cooperation on large scales is regulated by direct sanctions, not reciprocity. If ten thousand union members go on strike, free riding is not deterred by the threat that others will free ride in response. More direct sanctions are necessary. The same goes with many public goods whose provision is guaranteed by coercive institutions like police and courts.

A number of authors have doubted that direct sanctions, especially costly ones, play much role in cooperation in societies without such coercive institutions.[59] So let's consider the evidence from one well-studied case: Turkana warfare.

WHY THE TURKANA FIGHT

We cannot study how hunter-gatherer societies solve the collective-action problem inherent in warfare because no contemporary foraging group engages in large-scale conflict. However, there are contemporary groups like the Turkana of northern Kenya that do engage in warfare, and that are similar enough to foragers that they give us some picture of how it might have worked.

The Turkana are nomadic subsistence pastoralists who live in the arid savanna of East Africa. They make temporary camps and frequently move from one place to another. These settlements are larger than forager bands and include many unrelated individuals. There is no centralized authority or coercive institutions. Leadership roles are taken on by men who acquire prominence through their ability as diviners, their status as warriors, or their capacity to make wise migration decisions. While they play a role in coordinating action, these leaders have no recognized coercive authority. Nonetheless, the Turkana mount large-scale raids to capture cattle from members of other ethnolinguistic groups.

Our knowledge of Turkana warfare comes from a study by Sarah Mathew, who was once my graduate student at UCLA and is now my colleague at ASU. Sarah and I planned this research together, but she did all the real work. Mathew interviewed a representative sample of 118 Turkana men about the most recent raid they had participated in and obtained detailed accounts of eighty-eight different raids. She also collected information about people's reactions to hypothetical acts of bravery and cowardice and reconstructed the family structure of many men and women to estimate mortality from warfare.[60]

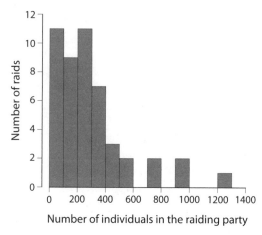

Figure 2.5. A histogram of the size of raiding parties in force raids in the Turkana data set. The mean group size is 315 men.

Turkana raids often involve many people. "Stealth" raids involve only a few men who aim to find poorly guarded livestock, but the Turkana also organize "force" raids that involve larger numbers of men who plan to engage in combat (fig. 2.5). On average, 315 men participate in a force raid.

Warriors experience substantial costs. A warrior who participates in a raid must leave his family and herds behind and may be injured or killed. Based on data from forty-seven force raids, Mathew estimates that warriors have a 1.1 percent chance of being killed on each raid. Data from Mathew's life history interviews indicate that 20 percent of all male deaths are due to warfare, 54 percent of these deaths occur when men are raiding other groups, and 46 percent occur when men are defending their own property against raids launched by members of neighboring ethnic groups.

These large-scale raids create collective benefits that flow to everyone in the war party and sometimes to everyone in the community. In successful raids, the war party acquires livestock that is divided among all the participants based on their age, not their contribution to the success of the raid. Some raids are also motivated by a desire for revenge against groups who have raided them in the past. Such raids produce deterrence that benefits everyone. For example, one raid was initiated very soon after a settlement was attacked so the enemy would know it was meant as retaliation. The warriors set out knowing they were not likely to acquire animals, and they considered the raid a success even though they returned empty-handed. Large-scale raids can also increase access to grazing areas and crucial dry-season water holes. One of the largest raids in Mathew's sample was precipitated by the discovery that herders from another ethnic group had settled at a watering site typically used by the Turkana. Mathew's informants said that their goal was to drive the intruders away from Turkana territory.

Turkana raids have all of the elements of a public good. They produce collective benefits at a substantial cost to individuals. Raiding parties are large, so the incremental effect of an individual warrior on the outcome is small, and the cost to individuals exceeds the benefit to them. So why do Turkana men participate? Why doesn't the whole system collapse under the weight of free riders, men who don't join raiding parties or shirk during battle?

And free riding does occur. A Turkana man is often faced with the option of joining a raid or staying back. Recruitment is informal. Raids are initiated by a few men who send out word to other settlements encouraging men to join a raiding party. A man can refuse to join, but only if he has a good reason. Once organized, the raiding party travels on foot for several days to an enemy settlement. Along the way, some men turn back. They may sneak away at night or tell their age-mates that they are ill or worried about their herds back home, or had a premonition that they will die on the raid. Such desertions occurred in at least 43 percent of force raids (fig. 2.6). Warriors also have many opportunities to reduce their risk during combat. The fighting typically begins when the raiding party surrounds an enemy settlement and opens fire. As soon as the offense gains the upper hand, fast young men begin to

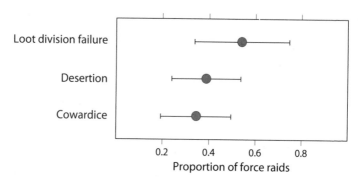

Figure 2.6. The proportion of raids in which different categories of free riding occurred.

drive the cattle toward Turkana land, while older warriors fight a rearguard action to slow pursuit. In 45 percent of force raids in which combat occurred, the warrior being interviewed knew of men who lagged behind others during combat, failed to fire their weapons, ran away when the firefight began, or retreated too quickly (fig. 2.6). The men driving the cattle homeward are supposed to continue until they are safely within Turkana territory and then wait for the rest of the warriors to rejoin them. Once reunited, they divide the spoils before they disperse to their respective settlements. Norms specify that members of senior age-groups get a larger allocation than members of junior age-groups, and that men within an age-group get roughly equal shares. However, the loot-sharing system failed in 56 percent of the force raids, and some participants took whatever livestock they were able to drive off (fig. 2.6).

Nonetheless, year after year the Turkana risk their lives fighting in large groups to produce collective benefits. What is it that allows them to solve this collective-action problem?

It's not kinship or reciprocity. Raiding parties are composed mainly of men who are not related and are not close associates. Turkana speakers are divided into twenty geographic groupings called territorial sections, each numbering around twenty thousand people. They also belong to

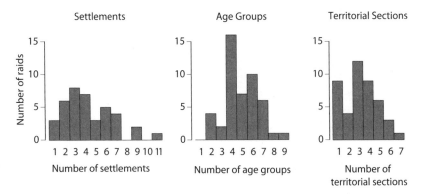

Figure 2.7. Histograms of the number of raids in which warriors were drawn from the number of settlements, age-groups, and territorial sections shown on the horizontal axis. Settlements are a group of people living together at any given time. They range in size from a few hundred to a thousand or so. Age-groups are made up of men who were circumcised at roughly the same time and number about five hundred. Age-groups are the most important social group for Turkana men. Membership in territorial sections is inherited and confers rights to graze in a particular region. They include about twenty thousand people.

patrilineal clans and age-groups that link men born within a five- to six-year period. While there are social ties that bridge these groupings, day-to-day interactions typically occur within them. The warriors participating in a raid are drawn from an average of five age-groups, four settlements, and three territorial sections (fig. 2.7). When asked whether they recognized the men gathered for a raid, participants typically responded that there were some men they knew, and some men they did not recognize.

Norms enforced by direct sanctions allow the Turkana to solve the collective-action problem in warfare. In 47 percent of the force raids in which desertions occurred, deserters were sanctioned, and in 67 percent of the force raids in which cowardice was reported, cowards were sanctioned (fig. 2.8). These sanctions are graded, ranging from verbal abuse to fines and corporal punishment. The latter can be

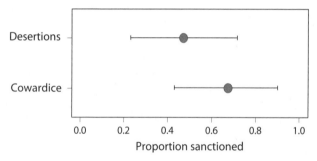

Figure 2.8. The proportion of different categories of free riding in which free riders were known to be sanctioned.

severe—the offender is tied to a tree and beaten by his age-mates. Community consensus determines how someone is sanctioned. Local people, especially age-mates, discuss the violator's behavior. Once a consensus emerges, members of the violator's age-group are responsible for administering punishment, even if they did not participate in the raid and did not experience the consequences of the violation.

It is very likely that positive incentives—rewards for men who are brave in combat—also play a crucial role in sustaining cooperation, but it is hard to know for sure. Rewards for bravery are not associated with a single act but instead accumulate over a lifetime. Measuring these diffuse benefits is difficult, and distinguishing the effect of behavior during warfare from the effect of other factors that contribute to a person's value as a social or mating partner is beyond the scope of Mathew's study. However, we can conclude from the rate of direct punishment that, important as indirect sanctions and rewards may be, they cannot be the full story: if they created sufficient incentives, there would be no need for direct punishment.

WHY PUNISH?

In small group interactions, there is an easy answer to this question. People punish to create incentives that motivate others to behave in ways that increase the punisher's fitness.

You punish a thief because he will think twice before stealing from you again. In this case, there is a direct causal connection between what acts you decide to punish and the effect on your own future payoffs.

There seems to be a big problem with punishment when large groups of people cooperate to provide public goods. Punishing is costly to the individual punisher and creates a benefit for everybody who benefits from the public good. Punishment may motivate contributions to a public good, but it creates another version of the same problem: Why punish? This is the infamous "second-order free-rider problem" and has been the subject of much puzzlement. Some authors think it means that punishment is unlikely to occur, while others explain punishment as a side effect of other motives, or the result of psychological machinery that evolved when people lived in small groups.

I think this puzzlement is misplaced. The solution to the second-order free-rider problem is the same as the solution to the first-order free-rider problem—norms enforced by third-party punishment. People punish because it is normative, and if they fail to punish, they risk sanctions. In models and laboratory experiments involving punishment, typically the only option is how much to punish. But in real life, punishment is embedded in a culturally transmitted system of moral norms that specifies what constitutes a punishable offense, what are mitigating circumstances, whether the evidence is sufficient, what level of punishment is appropriate, and who is obligated to administer the punishment. If you are responsible for punishing someone and don't do so, you have violated a norm. If you punish an innocent person, you have violated a norm. If you administer a harsher punishment than is justified, you have violated a norm. And if you violate a norm, you risk sanctions. So you are motivated to follow the norm.

We know this is true in complex societies in which punishment for many offenses is a state monopoly administered by coercive institutions. Judges who fail to administer the law because they have personal relationships with the accused or because the accused is rich or famous are morally reprehensible, and in societies with robust institutions they often end up in jail themselves. Sarah Mathew has conducted a study that suggests this is also true of the Turkana.[61] She read Turkana men and women very short stories, or vignettes, in which one warrior sees another warrior act in a cowardly way during a raid. In one vignette, the warrior who witnessed the act of cowardice tells the coward's age-mates about what he has done, and when a consensus is reached, the warrior participates in punishing the coward; in a second version of the story, the warrior who witnessed the act of cowardice does nothing and the coward is not punished. In each case Sarah asked participants what they thought about the warrior who witnessed the act of cowardice. Do you think he is wrong? Does he displease you? Was his action useful? Would you criticize him, punish him, help him, or marry him? As is shown in figure 2.9, the Turkana have positive attitudes toward this warrior when he punishes the coward and negative attitudes toward him when he doesn't. Sarah also told stories in which the first warrior punished too harshly or punished unjustly. The Turkana disapproved of these behaviors as strongly as they disapproved of a failure to punish.

Sometimes people say that this argument creates an infinite regress. Punishment creates a second-order free-rider problem that is solved by the punishment of nonpunishers. But this creates a third-order free-rider problem: Who punishes those who don't punish nonpunishers? Solving this problem creates a fourth-order free-rider problem, and so on forever. Keeping track of all this seems unreasonable to many people. I think there are two complementary answers to this argument. First, there is no need for actors to keep track of the entire sequence. They just need to know whether an actor violated a norm during the last period. If one of his age-mates failed to punish someone he was obligated to punish, there is no need to know why the punishment was appropriate.

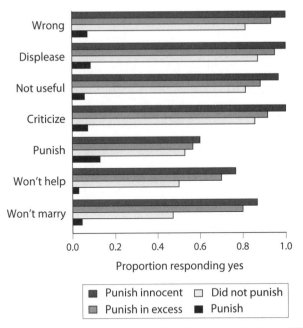

Figure 2.9. Turkana responses on hearing one of four different stories about an individual who sees a second individual behave in a cowardly manner on a raid. In the "punish" version of the story, the first individual punishes the coward after consultation with the community, while in the "did not punish" version of the story, he does nothing. In other versions, the first individual punishes unjustly, or too harshly. The horizontal axis gives the fraction of sixty participants who expressed the forms of disapproval listed on the vertical axis.

Or, more formally, in game-theoretic models, behavior can be conditioned on a small set of state variables that summarize past events and are updated each time period using only information about state and actions during that time period.[62] Real decisions about norm violations require lots of evidence about context, mitigating circumstances, and so on, and it is not obvious that keeping track of higher-order norm violations is actually any harder than keeping track of these things. Second, it is likely that violations become rarer as you progress from second- to third- to fourth-order punishment. In equilibrium, only a fraction of actors will fail to cooperate and require punishment. Then only a fraction of those people who were obligated to punish will fail to do so, and only a fraction of those people who were obligated to punish such nonpunishers will fail to do so. You can see what is going to happen—the expected cost of administering punishment diminishes with the product of these probabilities as we go from first- to second- to third-order free riding. This means that selection against nth-order free riding may be very weak, and therefore very weak evolutionary forces can stabilize the whole structure.

PICK AN ESS, ANY ESS

According to the argument I have made, the ultimate motivation for norm enforcement is that norm enforcement itself is normative. The psychology that causes people to internalize and enforce norms was favored by selection because people who failed to enforce their society's norms when they were responsible for doing so suffered disapproval and sanctions. I believe that living in societies in which norm violators suffered serious sanctions led to the genetic evolution of moral sentiments that caused people to be more cooperative, trusting, and willing to obey and enforce norms than other primates and thus less likely to suffer the costs of norm violation. There is much empirical evidence that people have moral sentiments not found in other primates, and that nowadays cooperation is sustained partly by such sentiments. I do not have space to discuss the evolution of moral sentiments here, but there are good discussions elsewhere.[63]

If people enforce norms because enforcement is normative, then it follows that the content of norms is not constrained by the enforcement mechanism. The same psychology can stabilize a vast range of norms. For example, the Turkana go to war in large groups; free riders are punished by agemates, and people who do not punish appropriately are also sanctioned. It's easy to see that as long as the cost of being punished is high enough, the behavior observed among Turkana is evolutionarily stable. But, and this is a big but, this fact doesn't explain Turkana behavior. Systems of rewards and punishments can stabilize a vast range of outcomes, including noncooperative ones. Mutually enforced sanctions could maintain the norm "You may steal your neighbor's cows to feed your family" or "You may not steal your neighbor's cows to feed your family." Similarly, punishment can maintain norms that are beneficial at different scales. "Do not steal a clan member's cattle, but the cattle of the rest are for brave men to steal" and "Do not steal the cattle of someone from your tribe, but the cattle of other tribes are for brave men to steal" are both group-beneficial norms, but one norm benefits clans and the other benefits tribes. The same goes for virtually all normative systems. Norms can specify that you are obligated to give a job to the best candidate, or they can specify that you must give the job to a fellow clan member. They can specify that you must marry your brother's widow, or that marrying your brother's widow is incest. The list goes on and on and on.

When a vast range of outcomes are evolutionarily stable, knowing that a norm is evolutionarily stable doesn't tell you very much. To make this more concrete, think about architecture. Suppose we start with a pile of bricks, mortar, and timber. The materials can be assembled to create lots of different brick buildings—a Romanesque cathedral like Milan's Sant'Ambrogio, a Georgian mansion like Monticello, a split-level ranch house like those that fill eastern American suburbs, or a Victorian confection like London's Saint Pancras station. Each of these buildings is stable under the laws of mechanics. Gravitational loads, wind loads, and various stresses interact to create a stable structure. So does knowing that Saint Pancras is a stable construction explain its structure? Obviously not. Stability under the laws of mechanics rules out structures like those that populate the worlds of Dr. Seuss, but a vast range of different structures are stable. Stability does not tell us why Saint Pancras is the way it is. The same goes for norms. Knowing which social arrangements are stable tells us what is possible but doesn't make any real prediction about what will actually occur.

Any real account of human cooperation (and in fact cultural evolution of normative systems more generally) ought to specify the processes that give rise to the norms that are actually observed—an "equilibrium selection mechanism" in the jargon of evolutionary game theory. Competition among culturally different groups is one such mechanism, and in the next section I will explain how this works and present evidence that this process has influenced the norms we see in the world. However, I also think that it is very likely that other equilibrium selection mechanisms are important, and I will speculate a bit about these further on.

CULTURAL GROUP SELECTION

In *The Origin of Species*, Darwin argued that natural selection will generate cumulative adaptation whenever three conditions are satisfied.

1. There must be a struggle for existence, so that not all individuals reproduce.
2. There must be variation so that some types of individuals reproduce at a higher rate than others.
3. Variation must be heritable so that offspring resemble their parents.

Darwin thought these three conditions also applied to human groups, and in chapter 5 of *The Descent of Man*, he argued that the distinctive features of human sociality resulted from selection among groups with different standards of morality. While most biologists revere Darwin, many (maybe most) reject the idea that selection among groups explains the evolution of cooperative behavior. The key problem with Darwin's idea is that selection among groups can be effective only if some process maintains sufficient heritable variation among groups, and this seems implausible to many scholars focused on the evolution of human behavior.

Pete Richerson, Joe Henrich, Sarah Mathew, and I have argued at length that Darwin was right when it comes to cultural evolution.[64] We have just seen that norms can create incentives that stabilize a vast range of behaviors. Often, these incentives can maintain differences in norms between neighboring groups even though people and ideas move between groups. This means that differences will be stable through time, and when groups split, daughter groups will resemble their parent. Neighboring groups often compete—militarily, economically, and for prestige. When they do, competition among groups can lead to the spread of some norms at the expense of others. In what follows, I will sketch the evidence that this process, which we term "cultural group selection," has played a role in shaping the kinds of culturally transmitted norms we observe in human societies. Once we have the evidence in hand, I will explain how this idea relates to the notion of group selection as it is commonly understood among evolutionary biologists interested in social evolution.

There is an immense amount of evidence that cultural transmission can create stable differences in norms between human societies. At the end of the nineteenth century, anthropologists fanned out across the world to study the peoples living in the world's colonial empires. A century of hard and sometimes dangerous work documented a dazzling array of social systems. Peoples that live only a few tens of kilometers apart can sometimes have different rules governing every aspect of social life, including whom you may marry, whom you consider kin, whom you must support in conflicts with others, what are crimes, how crimes are punished, and so on and on. Some of this variation is the result of environmental differences—patrilineal inheritance is more common among pastoralists than among small-scale horticulturists.[65] However, as we saw in the last chapter, much variation is due to common cultural descent, not environmental differences.[66] Many of the behaviors of coastal Salish groups are more similar to the behaviors of distant inland Salish groups than they are to the behaviors of their Wakashan neighbors who share the same coastal habitat. The reason Salish groups share norms governing a host of behaviors is that they share a recent cultural ancestor from whom they inherited norms. The utility of phylogenetic methods in studying culture tells us that this phenomenon is not limited to western Indians.

The norms and values that predominate in a group can affect whether a group survives, whether it expands, whether it is imitated by its neighbors, and whether it attracts immigrants. Norms causing a group to survive will become more common compared to those that lead to extinction. Similarly, norms that lead to group expansion or are more likely to be imitated or attract more immigrants will increase compared to those that don't.

Richard Sosis's study of two hundred communes in the United States during the nineteenth and twentieth centuries shows how cultural norms can affect group survival.[67] All of these communities eschewed private property in favor of communal production and consumption. Of these, eighty-eight were religious, and the rest were based on secular ideologies that stressed communal living. As shown in figure 2.10, communes based on religious ideology had much higher survival rates than communes based on secular ideologies. At "birth," only about half of the communes were religious, but almost all that survived for at least forty years were religious. This means that selection acts to increase the frequency of religiously based communes relative to secular ones. These communities also had different norms regarding the requirements for membership. Some made onerous demands on members—frequent prayer or celibacy, for example. Sosis found that religious communes with a larger number of costly requirements survived longer than those with fewer requirements (fig. 2.11).

These data illustrate how norms maintain the variation among groups. Different religious communes had very different norms. Some were relaxed, while others were demanding.

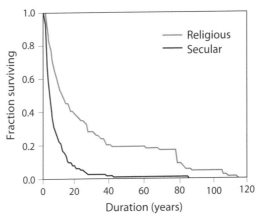

Figure 2.10. Proportion of communes surviving as a function of time since the founding of the commune. The black line represents 112 secular communes and the gray line 88 religious ones.

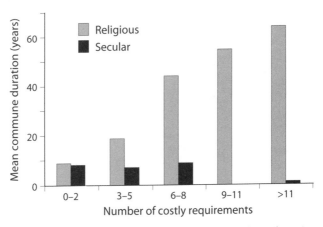

Figure 2.11. The mean duration of a commune plotted against the number of costly requirements that must be satisfied to maintain membership.

Why did people in the demanding communities put up with the costly requirements? Because the norms were enforced by group monitoring and the threat of ostracism and punishment. If you joined a commune like the Shakers that required sexual abstinence, you had to conform or leave. Different norms strongly affected the probability that the group would survive as a norm-enforcing unit. Only deeply committed people joined groups with many onerous requirements, and these groups survived the temptations produced by communal production and consumption longer than groups that attracted a wider range of people.

This result is consistent with recent cross-cultural experimental work. Joe Henrich and colleagues performed three experimental "games" in fifteen culturally diverse populations around the world.[68] For example, in the simplest, the "Dictator Game," one subject is offered the chance to divide a sum of money, keeping a fraction for herself and giving the rest to an anonymous second player. Belonging to a world religion like Islam or Christianity increased the fraction transferred by 23 percent across the fifteen societies. Henrich and colleagues argue that unlike the religions of many small-scale societies, world religions incorporate beliefs in gods who punish norm violators, making it less costly to enforce social norms, and such beliefs aided the evolution of the larger-scale societies of the last several thousand years.[69]

The spread of the Nuer at the expense of the Dinka in nineteenth-century Sudan provides an example of how differences in norms can affect group expansion. The Nuer and Dinka each consisted of a number of politically independent groups.[70] There were differences among the Nuer and Dinka in subsistence practices, norms surrounding marriage and bride-price, and political systems. The latter were especially important. A Dinka tribe was a group of people who lived together in a wet-season encampment, and the size of Dinka tribes was constrained by geography. Membership in Nuer tribes was based on kinship through the male line, allowing Nuer tribes to grow to three to four times the size of Dinka tribes. These differences gave the Nuer an advantage in conflicts with the Dinka because their tribes were larger and because warfare typically occurred during the dry season, when Nuer encampments were larger than those of the

Dinka. In the aftermath of these conflicts, about 130,000 Dinka were incorporated into Nuer polities and eventually assimilated Nuer norms.

This example illustrates the requirements for cultural group selection by intergroup competition. There must be persistent cultural differences between groups, and these differences must affect their competitive ability. Losing groups must be replaced by winning groups, but the losers need not be killed. They need only disperse or assimilate into victorious groups.

Such competition is common in small-scale societies. The best data come from New Guinea, which provides the only large sample of traditional societies that were studied by professional anthropologists before experiencing major changes due to contact with Europeans. Joseph Soltis assembled data about intergroup conflict from the reports of many early ethnographers in New Guinea.[71] Many studies report appreciable intergroup conflict, and about half mention cases of social extinction of local groups. Five studies contained enough information to estimate the rates of extinction in a population of neighboring groups (table 2.1). Typically, groups are weakened over a period of time by persistent conflict with neighbors and finally suffer a sharp defeat. When enough members become convinced that their group cannot resist further attacks, they take shelter with friends and relatives in other groups, and their own group becomes socially extinct. These data give us an upper bound on the rate of change due to cultural group selection. Suppose a new norm arose, and only groups without that norm went extinct. Then it would take about twenty generations, ..re hundred years, for the new norm to spread from one ..oup to most of the other local groups by cultural group selection. This might seem slow, but the history of the rise of ever larger and more complex societies in the Holocene has taken thousands of years.

A propensity to imitate the successful can also lead to the spread of group-beneficial cultural norms. People often

TABLE 2.1. EXTINCTION RATES FOR CULTURAL GROUPS
FROM FIVE REGIONS IN NEW GUINEA

Region	Number of groups	Number of social extinctions	Number of years	Groups extinct every 25 years (%)
Mae Enga	14	5	50	17.9
Maring	13	1	25	7.7
Mendi	9	3	50	16.6
Fore/ Usurufa	8–24	1	10	31.2–10.4
Tor	26	4	40	9.6

Source: Soltis et al. 1995.

know a lot about the norms in neighboring groups. They know that we can marry our cousins, but those people over there cannot; anyone is free to pick fruit here, while individuals own fruit trees there. Suppose different norms are common in neighboring groups, and one set of norms causes people to be more successful than other sets of norms. There is considerable evidence that people have a strong tendency to imitate the successful.[72] This means that beliefs can spread from successful groups to their less successful neighbors over a wide range of conditions.[73]

The rapid spread of Christianity in the Roman Empire may be an example of this process. In the 280 years between the death of Christ and the conversion of Constantine, the number of Christians increased from a small number to between six and thirty million people, a rate of increase of 3–4 percent per year. According to Rodney Stark, many Romans converted to Christianity because they were attracted to what they saw as a better quality of life in the early Christian community.[74] Pagan society had weak traditions of mutual aid, and the poor and sick often went without any help. Mutual aid was particularly important during the severe epidemics that struck the Roman Empire during the late Imperial period. Pagan Romans refused to help the sick or bury the dead. As a result, cities devolved into anarchy. In Christian communities, strong norms of mutual aid produced solicitous care of the sick and reduced mortality.

Selective migration can also lead to the spread of some kinds of group-beneficial norms. In the modern world, streams of migrants flow between societies. The extensive literature on this topic[75] supports two generalizations: (1) migrants flow from societies where their prospects are poor to ones where they believe their prospects will be better, and (2) most immigrant populations assimilate into the host culture within a few generations. Ethnographic evidence suggests that selective immigration is likely to be an ancient phenomenon.[76] The spread of cultural institutions associated with ancient complex societies, such as China, Rome, and India, suggests that this process is not new. Ancient imperial systems often expanded militarily, but the durable ones, like Rome, succeeded partly by inducing an inflow of migrants who then assimilated Roman norms. Although the Roman Empire as a political entity eventually faded, its most attractive institutions were adapted by successor polities and persist in modified form to this day.

Let me be clear. Cultural group selection does not lead to progress. It leads to the spread of norms that make groups successful in competition with other groups. Since the origin of food production about ten thousand years ago, norms that allow larger groups to cooperate have gradually spread in most parts of the world. Whether this represents progress is a matter of opinion. The resulting increase in social scale led to wealth, long life, security, and intellectual delight for some people at some times in some places. But in other times and places, it produced hierarchy, oppression, slavery, and genocide. Before the origin of agriculture, there were at least one hundred thousand years of cultural evolution in which norms leading to smaller, poorer, but more egalitarian societies were favored. Moreover, cultural group selection is a relatively slow process that depends on competition between long-lived entities. In a contemporary world in which external conditions are changing at a frantic pace, there is no reason to suppose that existing norms are well suited to current environments.

WHY MANY PEOPLE REJECT CULTURAL GROUP SELECTION

I hope you will agree that the idea of cultural group selection is plausible. It fits with what we know about human history, and there are lots of examples. Also, to the extent that norms regulate behavior in firms, universities, and other institutions within complex societies, they can play a role in the evolution of these institutions as well. For example, it is a truism that business firms vary culturally. Some have norms that promote freewheeling innovation, while others are conservative and bureaucratic. If so, cultural group selection will shape the kinds of firms in the world as long as these differences affect the extent to which firms survive, grow, and are imitated by other firms. The same goes for religious institutions, clubs, political parties, and a variety of other culturally variable groups within complex societies.

If this all seems reasonable, you may be surprised to learn that cultural group selection is controversial among students of human evolution. Among biologists there is vigorous debate about whether it is useful to think about genetic selection among groups. The population genetics theory of selection among groups is very well worked out. Most of this theory focuses on behaviors that are beneficial to groups but costly to individuals, and the central question is: Can selection among groups overcome selection within groups? To a first approximation, the answer to this question depends on the fraction of genetic variation among groups.[77] Increasing the amount of genetic variation among groups makes selection among groups more important in determining the net direction of selection, and if there is enough variation among groups, selection among groups will prevail. It seems pretty simple.

The debate arises because you can also describe these genetic group selection models in terms of inclusive fitness (or synonymously, kin selection). To get a sense for the tone of this argument, look at Steve Pinker's "The False Allure of Group Selection" and associated commentary on the Edge website: http://edge.org/conversation/the-false-allure-of-group-selection. When most of the variation is among groups, relatedness within groups is high, so inclusive fitness maximization also predicts that group-beneficial behaviors are more likely to evolve. The two approaches are equivalent ways of doing the accounting for the same underlying process. If you do your sums correctly, both approaches will produce the same answer. My view is that these approaches are both useful, and many biologists agree.[78] However, there are also many who believe that it is *always* confusing and misleading to take the group selection approach since it is really just kin selection.[79] Most scholars interested in the evolution of human behavior learned their evolutionary biology from people who take this view, so the received view in this community is that genetic group selection is never important.

Cultural group selection is different because it is about selection among groups with different social arrangements that are evolutionarily stable, *not* about the evolution of individually costly group-beneficial behavior, and this makes all the difference. As we have seen, norms can stabilize a vast range of different behaviors, and because cultural adaptation is often fast compared to migration and other forms of mixing, adaptive cultural learning processes can cause norms to stabilize very different behaviors in different groups.[80] This means that within each group, behavior is determined by self-interest. But there are a vast number of social arrangements of this kind, each one consistent with individual self-interest. When this is the case, competition between groups may often determine the social arrangements we observe in the world.

Note that variation among groups is still important, but now it is created and maintained by adaptive cultural learning processes, the analogues of selection in cultural evolution. Natural selection does not usually maintain genetic variation among social groups because it is usually too weak to counteract the effect of migration, and in genetic models of social evolution, variation among groups is maintained by random genetic drift. This difference has two important consequences. First, in genetic models the amount of variation among groups is sensitive to group size. All other things being equal, as group size increases, relatedness declines.[81] In cultural models, variation among groups does not depend on group size because cultural learning processes can create and maintain lots of cultural variation among very large groups.[82] Second, when there are multiple equilibria and when adaptive processes are strong compared to migration, groups can stabilize at different frequencies, and knowing the frequencies of traits in the population as a whole does not allow you to predict the amount of relatedness within groups. You need to actually keep track of the dynamics of the frequencies in each group.

As a result, it is difficult to apply inclusive fitness to analyze cultural change when adaptive processes are strong enough to maintain variation among groups, and to my knowledge, no one has successfully done this kind of analysis. In fact, Stuart West and colleagues[83] argue for models in which adaptive processes are weak and cultural variation among groups is maintained by the cultural analogue of genetic drift because this allows the use of standard inclusive fitness methods. The only problem is that such models do not fit the empirical patterns of cultural variation.[84] In contrast, the group selection approach has been successfully used for more than thirty years to model equilibrium selection due to intergroup competition.[85] Inclusive fitness and group selection approaches are equivalent, so we should use the approach that is most useful. In the present case this seems to be the group selection approach.

If I had to do it all over again, I would not use the term "cultural group selection," because it triggers a negative reaction from those who think that group selection is relevant only when there is conflict between the individual and the group. I did not anticipate this response when Pete Richerson and I first developed this idea in the early 1980s. At the time, I was a postdoc in Michael Wade's lab at the University of Chicago. Mike was one of the foremost proponents of the idea that selection among groups is an important process, and I was surrounded by people who shared this perspective. Moreover, there was a lot of excitement about Sewall Wright's idea that group selection plays an important role in genetic evolution when interactions between different genes create multiple evolutionary equilibria. Mathematical models of the process that Wright envisioned are very similar to some models of cultural group selection.[86] Now, thirty years of arguments later, I can see that it would have been wiser to use the term suggested by Sarah Mathew and Matt Zefferman, "group-structured cultural selection."[87]

CULTURAL GROUP SELECTION
CAN'T BE THE WHOLE STORY

I think cultural group selection is a plausible mechanism, and there is much evidence to support the idea that it is one of the processes that shape norm content. However, it also seems to me that there are many clear examples of norm shifts that cannot be explained as a consequence of group competition. The shift in norms regarding dueling provides a good example.[88] Up until the first part of the nineteenth century, violations of honor norms required upper-class European men to challenge one another to duels. Thus, in 1829 the sitting prime minister of the United Kingdom, the Duke of Wellington, challenged the Earl of Winchilsea to a duel because the earl had accused him of dishonesty in the parliamentary debate over the Catholic Emancipation Bill. (It is as if President Obama felt obligated to risk his life defending his honor against Ted Cruz.) As it turned out, Wellington, a notoriously poor shot, missed, and Winchilsea then discharged his pistol into the air. Twenty-five years later, dueling had completely disappeared from British society. In the United States over the last few decades there have been shifts in norms about smoking, premarital sex and giving birth to children out of wedlock, and same-sex marriage. It seems implausible that any of these shifts were caused by competition between groups. There has been no group extinction or spread, and we did not copy the new norms from anybody else.

We don't have a very good theory explaining such shifts in norms. Historians provide plausible narratives for particular cases. For example, Kwame Appiah[89] argues that dueling disappeared as the result of the increasing importance of commercial and manufacturing interests among British elites. Game theorists[90] offer models similar to "peak shift" models in population genetics,[91] in which statistical fluctuations due to finite population size cause random walks in trait frequencies. A population at one equilibrium eventually randomly walks to an alternative equilibrium, stays a while, and then randomly walks back again. I don't think that these models fit the facts. First, the waiting times for norm shifts are extremely long in large populations, especially if there are substantial costs to deviating from existing norms. Second, they are reversible. Shifts in one direction may be more probable than shifts in the opposite direction, but eventually every population will shift back and forth many times. Changes in the model that make this less likely (bigger populations, lower error rates) also increase the waiting times until a shift occurs.

NORMS AND SMALL-SCALE COOPERATION

So far I have concentrated on large-scale cooperation. However, humans are also exceptional cooperators at smaller scales. Division of labor and delayed exchange occur in every human society, and as we saw earlier, exchange plays a key role in the economics of foraging societies, leading to lower mortality and shorter birth intervals than in other apes.[92] There is much less small-scale cooperation in other mammals, and the cooperation among unrelated individuals seems limited to low-cost behaviors like grooming in non-human primates.[93]

This fact suggests that reciprocity alone does not explain small-scale cooperation in humans. Many other vertebrates live in stable social groups and can remember the past behavior of others and adjust their behavior accordingly. If it is reciprocity alone that stabilizes exchange in human societies, then we should see lots of exchange in these species. But we do not. Moreover, the fact that small-scale cooperation in other mammals is limited to low-cost behaviors also suggests that reciprocity is not at work. Models of reciprocity predict that it is not the cost that is crucial; rather it is the ratio of benefits to costs that matters. This means that reciprocity can support food sharing that imposes substantial costs on donors as long as it generates even larger benefits for recipients. For example, when the recipient is injured and cannot feed herself or is exhausted after giving birth, benefits may be much larger than costs. Since these are common events, aid should be commonplace. Yet humans are one of the few mammals that regularly share food, and when we do, the benefits are large.[94]

In what follows I am going to argue that norms allow humans to organize more small-scale cooperation than other animals because norms solve the problems of monitoring and enforcement that limit the evolution of reciprocity in other species. This idea originated with Sarah Mathew, and, in collaboration with Matthijs van Veelen, we have explored it at greater length elsewhere.[95]

Norms Regulate Human Small-Scale Cooperation

Accounts of social life given by evolutionary thinkers often have a libertarian flavor. Society is conceptualized as a network of bilateral bargains between self-interested individuals and nepotistic families. Sanctions, and therefore morality, arise because actors seek to modify the behavior of others in ways that increase their own welfare. Some authors would concede that norms are necessary for large-scale cooperation, but most would say that they have no role in explaining small-scale cooperation.

However, this picture doesn't fit the facts. Small-scale cooperation in human societies is typically regulated by shared norms that are enforced by third-party sanctions. Individuals are not free to make any bargain they want; deals are constrained by existing norms. Moreover, these norms affect behavior that is closely connected to fitness, like pair-bonding and parenting. Most societies recognize "marriage," an institutionalized form of pair-bonding, and marriage is associated with normative rights and obligations. There are often norms that forbid pair-bonding between members of the same social unit and require pair-bonding between members of different social units, and these norms usually trump the interests of potential marriage partners. Among the Walbiri of Australia's Western Desert, people belonged to one of eight sections. Members of each section were supposed to marry members of only one of the other seven.[96] Norms specify how many people a person can marry. A man and two women cannot choose to marry in a society that is normatively monogamous even if they think it is in their mutual long-term best interest. Norms regulate the direction of wealth transfer at the time of marriage, so that in some societies men provide payment to the bride's father and in others the bride's family is expected to provide payment to the groom's family. Postmarital residence is often regulated by norms. A man who hails from a patrilocal culture cannot decide to live with his wife's family without losing face.

Norms also regulate how individuals raise their offspring. In state-level societies, norms regulate how you discipline your children and how you educate them. There is much cultural variation in these norms, ranging from places where it is illegal for parents to beat their children to ones where parents who do not do so are considered negligent. Sending daughters to school was counternormative in many societies until recently, causing a substantial fraction of women to be excluded from the market labor force. Parental investment is regulated by inheritance norms that specify how wealth should be distributed among children—only to sons or only to daughters, to the oldest, to the youngest, or equally divided among all the children. It would be wrong to practice primogeniture in a society where norms specify even division or to deny daughters an inheritance in societies where all children should be given a share.

Interaction between long-standing friends seems like the canonical case of direct reciprocity. Despite this, community approval and disapproval provide vital scaffolding in sustaining the cooperation between friends among the Turkana.[97] For instance, a woman who refuses to give water to a thirsty man can be criticized. However, if she heard that this man had abandoned his injured friend when the two of them went into enemy territory to steal cattle, then she can refuse him water without facing disapproval. Similarly, a herdsman can expect to be hosted by Turkana households when he travels in search of stray livestock. But if word has it that this man has stolen his neighbor's camel, he may be refused hospitality. The relationship between neighbors is regulated by community sanctions, not just their own history of interactions.

Norms also come into play when there are disagreements within pairs of individuals. If a Turkana man's goat is stolen, he reports the matter to the elders or members of the respective families. They summon the alleged thief to determine what happened, and if he is found to be at fault they instruct him to compensate the victim.

Don't mistake what I am saying here. People are not norm-following robots. They are actors with their own interests, and they make deals with others to further these interests. However, norms affect the kinds of deals that people make. First, people have to take into account the cost of violating norms. In a society in which sexual fidelity is highly valued, potential marriage partners cannot agree to an open marriage without suffering the consequences. In a society in which polygyny is thought to be immoral, a poor woman cannot choose to be the second wife of a rich man without suffering serious costs—think of the response to Mormon plural marriage in nineteenth-century Illinois. Second, and I think more important, people internalize their society's norms, and this affects their preferences and thus the choices they make. In a society like ours in which corporal punishment of children is counternormative, most people believe that corporal punishment produces bad outcomes and feel shame if they lose their temper and hit their child. In other societies, people internalize alternative beliefs about how to discipline their children and may feel shame when they refrain from punishment.[98]

HOW NORMS STABILIZE SMALL-SCALE COOPERATION

So far the argument has been empirical—humans exhibit many forms of small-scale cooperation not seen in other species, like specialization and exchange, and shared norms play a crucial role in regulating many forms of small-scale cooperation in human societies. There are also good theoretical reasons to think that norms and third-party enforcement can extend the range of conditions that favor small-scale reciprocity. Let's consider several of them.

When Perception Errors Are Common, Third-Party Monitoring and Enforcement Help Stabilize Reciprocity

Standard models of the evolution of reciprocity predict that reciprocity should be very common in nature. According to these models, contingent strategies like Tit-for-Tat can support sustained cooperation among unrelated individuals when

$$\left(1 - \frac{1}{T}\right)b - c > 0$$

where T is the average number of interactions between reciprocators and b and c are the incremental effects of cooperation on the recipient's and donor's fitnesses.[99] Note that the left-hand side of this inequality is like inclusive fitness (defined above) except that the incremental effect of a behavior on the fitness of the recipient is weighted by $(1 - 1/T)$ instead of relatedness. To see what this means, assume that individuals interact only twice on average. Then according to the standard models of reciprocity, contingent cooperation is stable as long as the benefits exceed twice the cost—the same benefit-cost ratio that favors cooperation among full siblings. If there are ten interactions, benefits have to be only 10 percent more than costs. Thus, according to the standard theory, reciprocity should be very common and should lead to cooperation with lower benefit-cost ratios than we see among brothers and sisters. Obviously, this is wildly wrong. Relatedness explains lots of costly cooperation in nature, and reciprocity very little.

The standard theory has many unrealistic assumptions that might explain the discrepancy between theory and fact. I think the most likely culprit is that the standard theory minimizes the effect of errors. In most models there are two options: cooperate or defect. In real life, there are many possible interactions with different costs and benefits. A reciprocator has to determine whether his partner's behavior was a real defection (she decided to ski for an extra day instead of meeting him at the airport) or an error (she meant to meet him, but her car got stuck in the snow and she couldn't make it). This means that it is easy for reciprocating individuals to disagree about whether cooperation or defection occurred. Sometimes individuals will think they cooperated, and their partner will disagree. Such misunderstandings are labeled "perception errors," and if they are common they can severely limit the possibility of reciprocity. The standard remedy is a special strategy called "win-stay-lose-shift," which recovers from perception errors.[100] However, unlike many other reciprocating strategies, win-stay-lose-shift can't increase when rare because it is easily exploited by defecting strategies. Moreover, people don't show even the slightest hint of following this strategy in experiments in which perception errors occur.[101]

Third-party monitoring and punishment can increase the stability of reciprocity when perception errors are common by reducing the error rate and increasing the penalty for defection. Suppose members of the community monitor the behavior of others and evaluate their conformance to norms. These observations are pooled and a consensus is reached. Later decisions about whether an individual receives help are based on the community consensus. Individuals who don't participate are punished by withdrawal of future help, thus eliminating the second-order free-rider problem.[102] Such third-party monitoring and punishment allows reciprocity to persist under a wider range of conditions for two reasons. First, increasing the number of individuals who monitor behavior increases the chance that real defectors will be exposed and decreases the chance that honest mistakes will be falsely labeled as defections. Second, increasing the number of people who punish defection can make it less attractive to defect. Modeling work suggests that these effects can be very substantial.[103]

Adjudication by third parties can also increase the stability of reciprocity by providing a mechanism that allows individuals to reestablish cooperation after a misunderstanding. Suppose I think that my friend has stood me up at the airport, but she really got stuck in the snow. My friend thinks that she has cooperated, but I think she has defected, and I refuse to help her the next time she asks. My friend, thinking she did nothing wrong, retaliates in turn, setting off a series of reprisals. However, now suppose that third parties adjudicate the initial dispute and decide whether my friend actually defected. This provides a clear public signal, and by conditioning behavior on that signal, individuals can reestablish mutually beneficial long-term cooperation. Modeling suggests that adjudication can greatly increase the range of conditions under which reciprocity can evolve, because it aligns the beliefs of interacting individuals even if the adjudication process is also error prone.[104] Adjudication of pair-wise exchange is easiest if rules of behavior are shared within a community, and not restricted to a particular partnership. Otherwise, third parties have to evaluate deviations from the pair-specific rules when arbitrating disputes.

Norms Can Help Identify How to Cooperate in the Local Ecology

Humans live in a wide range of environments with different subsistence systems, residence patterns, and social arrangements. This means that the actual strategies that govern reciprocity in different environments will need to be learned. For example, hunter-gatherers frequently share meat, but this is not a simple problem of dividing a homogeneous resource. Different parts of the animal have different value (more lipids vs. more protein) to different people (old vs. young, male vs. female). Reciprocity seems to be hard to learn using general-purpose learning mechanisms. A number of authors have attempted to teach other animals to behave reciprocally in simplified laboratory settings, but without much success.[105] People are definitely better learners than other animals but still find it difficult to coordinate on cooperative strategies even when there are no errors and the cooperation is an ESS.[106] In real-world environments with lots of contingencies, it seems likely that it would be even harder for a population to converge on a mutually sustainable strategy.

Culturally evolved norms of reciprocity can help solve this problem. Reciprocity does play a role in meat sharing among foragers,[107] but it is often regulated by detailed cultural norms that specify which individuals get what part of the animal. For example, among the Mbendjele, a group of Congo Basin foragers, norms specify that the hunter receives the heart, other men the kidneys, and the hunting dog the lungs, and that the rest should be shared equally lest the hunter's luck be spoiled.[108] Such culturally evolved norms provide a scaffold to guide individual sharing decisions and limit the scope for conflict. An ethnographer once told me a story that fits with this picture. He had the practice of handing out small amounts of loose tobacco to people in the group he studied. He found this task annoying because there was lots of nagging and wheedling, so he asked a local man if he could divide the tobacco. The man declined, saying in urgent tones, "Don't do that, we will kill each other." These were foragers who shared meat almost every day. One interpretation of this anecdote is that they could share meat because they had culturally evolved norms about who deserved what in any given context, but they despaired of peacefully dividing a valuable but novel resource.

Norms Can Enforce Pair-Wise Exchange Beneficial to the Group but Not the Dyad

For instance, parents have a greater interest in cooperation between their children than do the children themselves. Imagine that two sisters are well placed to help each other when times get bad. If one sister gets hurt, the other can feed her. Inclusive fitness predicts that they will help as long as the marginal benefit of a transfer to the sister receiving the help is more than twice the cost to the helper. The parents, however, have an interest in their offspring being more helpful than that, and if they could, they would bind their children to help whenever the marginal benefit exceeded the marginal cost. Such conflicts of interest lead to a tug-of-war between parents and offspring, and the outcome is difficult to predict. However, a norm that siblings should help each other could work even better if it was shared by a larger community, as on average everyone is better off living in a group in which siblings help each other whenever it is mutually profitable. Within groups, people obey norms because violators are sanctioned, and under the right circumstances, group-beneficial norms could spread because they make the group more competitive.

HOW NORM PSYCHOLOGY MIGHT HAVE EVOLVED

Sarah Mathew and I think that both large-scale cooperation and small-scale cooperation in humans are regulated by shared, culturally evolved norms enforced by third-party monitoring and sanctioning. In contrast, many evolutionary thinkers believe that small-scale cooperation evolved because of the effects of genetic relatedness and direct reciprocity, and therefore the only thing that matters is the effect on me and my family. My friends Joe and Moe have made a deal that seems unfair to me. But why should I care? And more importantly, why should I make any effort to do something about it? My neighbor Alice beats her child, but this should be no skin off my nose. But real people do care. Self-interest alone doesn't explain how people operate in real human societies—we judge norm violators and often do something about our judgments. Of course, self-interest is always important. But so, too, are the social norms that shape every aspect of our social lives. Why should this be so?

We think that norms and the sanctioning of norm violators may have arisen to support a novel form of cooperative breeding in humans. According to this hypothesis, the capacities that allow cumulative cultural evolution evolved for other adaptive reasons—such as learning how to make complex tools or process foods to remove toxins. Norms developed about the best way to do certain tasks. Culturally transmitted information included rules about social interaction, for example the best way to structure meat sharing. Then in relatively small groups, the benefits associated with third-party monitoring and punishment led to the evolution of a norm psychology that allowed for more extensive small-scale cooperation in early human societies and may have helped weakly related bands to seize benefits from social exchange.

This led to the evolution of a moral psychology that structured the subsequent evolution of larger-scale cooperation through cultural group selection. Cultural group selection models have typically assumed that individuals acquire complex normative behavior by copying successful or prevalent behavior—that is, they can use the same social learning machinery that they use to learn other kinds of locally adaptive behavior to acquire local moral rules. But complex moral behavior may be hard to acquire without some kind of innate scaffolding already in place. Many of our normative concerns are somewhat abstract, as we recognize similarities in situations where there are aligned and opposed interests. That gives our moral machinery a complex structure. We recognize common causes and conflicts of interest in novel situations, and we link them to our norms. It may be difficult to maintain such rules with only a general cultural learning mechanism. But if small-scale cooperation maintained by norms and third-party enforcement favored the evolution of innate moral intuitions, then cultural group selection could lead to large-scale cooperation more easily by making use of this moral psychology.

This hypothesis is also consistent with the observation that different kinds of norms regulate small- and large-scale cooperation. Norm compliance and enforcement in small-scale cooperation should depend on cues of family membership or cues of past interactions. Such cues are rooted in individual identities. Should I help Joe? Is he my relative? Has he helped me when I needed him in the past? Negotiation, deliberation, and consensus among known individuals should be important in achieving norm compliance in small-scale cooperation. In contrast, cultural group selection on institutional variation should be more important in shaping norms governing large-scale cooperation. Norm compliance in large-scale settings should depend on cues of group identity. Should I help Joan? Is she a member of my ethnic group? Is she an American? In such settings, norm compliance can be achieved even without negotiation and consensus.

CONCLUSION: FIVE THOUSAND FENCE POSTS AND NARDOO CAKES

Let's rewind the tape of human evolution back four million years. Except for a peculiar way of getting around on the ground, our ancestors were a lot like the other apes around them. They ranged across much of sub-Saharan Africa, but not beyond, and lived in groups that were much like those of the other apes. They had no large-scale cooperation and very little specialization and exchange. Now, fast-forward to the beginning of the Holocene, ten thousand years ago. Modern humans have become outliers in the natural world. We have expanded across the globe. We thrive in a wider range of habitats, exploit a wider range of foods, and form more diverse kinds of groups than any other mammal. Humans have also become much more cooperative than any other mammal. On small scales, we relied on specialization and exchange, and this enabled us to rear many slow-growing, expensive children. On larger scales, we worked together to produce public goods benefiting many people. Fast-forward to the present, and here we are trying to figure out how and why this remarkable transition occurred.

My thesis is that the evolution of cultural adaptation was an essential ingredient in both our ecological success and our ability to cooperate. It wasn't the only factor, but it was a crucial one. We spread across the globe because we were able to culturally evolve adaptations to local conditions. The ability to make nardoo cakes and verbine twine nets was essential in central Australia, and the ability to make caribou-skin clothing and seal-oil lamps was equally necessary in the high Arctic.[109] No other creature is able to create so many different local adaptations. People are probably smarter than other creatures, but the complex local adaptations we depend on are beyond the inventive capacities of individuals. They are the result of a gradual cultural accumulation of knowledge. But this changes everything, because the ability to accumulate cultural knowledge comes with a built-in trade-off. Cultural accumulation works best when people are inherently motivated to adopt the beliefs of people around them, even when those beliefs conflict with their own experience. And this makes people partly creatures of their local culture. They believe the things they believe, and do the things they do, in part because that is what people around them believe and do. And this, in turn, opens the door for social norms to influence human behavior. Large-scale cooperation among unrelated individuals seems to work only when the actors' behavior is influenced by norms enforced by third-party sanctions. To get hundreds of people to work together to build a drive line with five thousand fence posts, there must be incentives to make their effort worthwhile. The same goes for building fish weirs and irrigation projects, and, probably most important, battling neighboring tribes. We have to share norms about what behavior is required, who imposes sanctions, how guilt is determined, and what level of sanctioning is appropriate. We don't know for sure how the psychology that makes all this possible evolved, but it is clear that shared beliefs about normative behavior are essential.

Several years ago I was invited to a conference where I was asked to speak on a topic titled "Does Culture Override Biology?" When I gave the talk, I changed the title because culture can't override biology. It can't because culture is as much a part of human biology as our peculiar pelvis and the thick enamel that covers our molars. Four million years ago, culture likely played a minor role in the life of our ancestors. Today, we are culture-saturated creatures, completely dependent on information acquired from others. Culture has allowed us to evolve highly refined adaptations that completely changed human subsistence and life history. It also allowed us to evolve complex, highly cooperative social arrangements unlike those of any other creature. But there is nothing unnatural or nonbiological about any of this. The morphological, physiological, and psychological changes that make human culture possible evolved over the last several million years as a consequence of the usual evolutionary processes. Culture has made us a very different kind of animal, but without doubt, we are still part of the natural world.

COMMENTS

CHAPTER 3

IMITATION, HAYEK, AND THE SIGNIFICANCE OF CULTURAL LEARNING

H. Allen Orr

I am neither an anthropologist nor a biologist who works directly on human evolution. Instead, I am a general evolutionary biologist who studies speciation and adaptation. My perspective on human cultural evolution is, then, necessarily that of something of an outsider. I hope, however, that my outsider perspective may provide some advantages along with its more obvious disadvantages. In particular, it may be useful to have some sense of how an evolutionary analysis of human beings resembles and (perhaps necessarily) differs from analyses of other species.

I find much to agree with in Professor Boyd's views. Perhaps most important, I agree with him that human beings are close to an outlier in the biological order. Human beings do live in essentially every ecosystem on the planet, and we do thrive in ways that other species could only envy—if they could envy. With the emergence of human beings it seems to me that something close to a phase transition occurred in evolution. The reluctance of some biologists to acknowledge this exceptional status seems to me odd. It is perhaps a relic of that Victorian program, led by T. H. Huxley and others, that subtly (and sometimes not so subtly) demoted human beings at every turn by reminding us that we are cousins to katydids. This claim is of course true, but claims can be both true and beside the point.

And the point now—now that there is no serious doubt about our biological nature—is, why are we so unusual among species? It is important to recognize that this question is nontrivial only *because* there is no doubt about our biological nature. Before Darwin, the answer to this question was trivially obvious—because we are nearer to angels than to animals. After Darwin, that answer no longer sufficed, and the question grew more serious, not less.

I find Boyd's approach to this question appealing for a number of reasons. First, his approach acknowledges the roles of biology *and* culture in human evolution, paying more than lip service to each. He also acknowledges the power of Darwin's idea of natural selection, whether that selection acts at the level of genes or also at the level of culture. I should stress, incidentally, that I have no principled problem with the notion of a Darwinian evolution of human culture. It has been clear formally, from work dating at least to George Price and his Price Equation (1970, 1972), that natural selection can operate simultaneously at multiple levels in the biological hierarchy: genes, organisms, species, and even cultures. If evolutionary biologists hesitate sometimes to invoke higher-level selection, it is not because the idea is deemed incoherent or is particularly difficult to grasp. It is because hard empirical evidence for natural selection that acts at higher biological levels is often lacking or is open to plausible alternative interpretations. But that topic, a difficult and technical one, is not one that we need consider in detail here (for my own view, see Orr 2015). For present purposes, it will suffice to note that (1) natural selection can, in principle, act at any level in the biological hierarchy; and (2) even if the evolution of human culture is not identical in form to the evolution of genes or organisms, as Boyd seems to recognize, it is perhaps similar enough to do some work, to get us somewhere. (There have, of course, been other important approaches to the Darwinian evolution of culture, including that of Cavalli-Sforza and Feldman [e.g., 1981] on gene-culture coevolution.)

I also find Boyd's approach appealing because it is grounded in explicit mathematical models of the sort that are familiar to—and that have proved invaluable to—population geneticists. To point to one example from his discussion, consider Boyd's summary of one of his many formal analyses, that of Perrault, Moya, and Boyd (2012). They consider the problem of identifying the circumstances under which an individual should imitate the behavior of others versus trust his or her own inferences about the state of an environment when environmental cues are noisy and thus unreliable. Their conclusion, the result of formal mathematical analysis, is that an individual should imitate the behavior of others when the environment changes slowly and/or when environmental cues are noisy. This type of analysis seems to me exactly the right way to begin to address such questions.

It is not the way, of course, to finally *answer* these questions. That will take a large body of rigorous and, I hope, clever experiments. And (again, from my outsider perspective) a reasonable case might be made that the field has lagged in performing enough of these experiments. But if the history of population genetics is any guide, the formal theory typically comes first, with a considerable lag before the appropriate experiments are performed. More to the point, it often isn't even clear what sorts of experiments are needed until theorists have elaborated a reasonably mature theory. (As an example, almost the entire corpus of classical experimental population genetics, including the efforts of Theodosius Dobzhansky to discriminate between balancing selection and processes like mutation-selection balance, rests on the prior efforts of mathematical theorists, including Sewall Wright, Ronald A. Fisher, and J. B. S. Haldane.)

Finally, and perhaps most important, I find Boyd's approach appealing because it does not push a single idea further than it can go. By contrast, while I believe that some truths may well reside in evolutionary psychology or even in the idea of memes, proponents of these schools seem to succumb sometimes to a temptation to push a single idea too hard, too far.

Despite all this, one of the things that every scientist learns sooner or later is that the set of appealing theories and the set of correct theories are not necessarily identical. (Indeed, they sometimes show disconcertingly little overlap.) So let me turn from these largely methodological matters to two concerns with Boyd's thesis. Neither, as we will see, rises to the level of serious challenge.

BIG BRAINS VERSUS HIGH-FIDELITY IMITATION

Boyd's main claim is that what makes human beings special is our capacity for cultural learning. To summarize, he argues that given the kinds of environmental changes that human beings have encountered through both space and time, traditional genetic adaptation is often too slow to explain our flourishing, and individual learning is often asking too much.

I agree with him on the first point. Although adaptation via genetic change may be faster than many evolutionary biologists once believed (see, for instance, Lescak et al. 2015), and although such adaptation surely played some important part in the success of anatomically modern humans, it is implausible to think that it accounts for the rapid spread of human beings into virtually every niche on the planet, in some cases very recently. With respect to the second point—the limited power of individual learning—Boyd's case of the lost European explorer seems to me fairly persuasive. If the prowess of the individual mind explains how human beings have flourished on Earth, then the European explorer who finds himself lost in an alien and hostile environment should also flourish. But, often enough, he doesn't. His individual cognitive powers are, sadly, frequently not up to those tasks—say, net making—that are performed routinely by the indigenous peoples who surround him.

In the end, Boyd spends much of his time contrasting these two possible cognitive explanations of our unusual status as a species. The traditional one is, roughly, that we are so smart, that we have such big brains. As evolutionary psychologists sometimes put it, our superior smarts have allowed *Homo sapiens* to invade an open "cognitive niche" (Tooby and DeVore 1987). The second explanation, and the one that Boyd champions, is that human beings are such good imitators of the behavior of others.

One of my concerns with Boyd's view is that it is easy to exaggerate the difference between these explanations. The point is that successful imitation of the behavior of others requires considerable brainpower. Our own culture places such a large premium on innovation that it is perhaps easy to underestimate just how much neuronal firepower is required merely to imitate successfully. This firepower is perhaps not the stuff of the Stanford-Binet Intelligence Scale or SAT scores. Nonetheless, high-fidelity imitation requires considerable social intelligence, subtle cue reading, and the solving of daunting frame problems. Boyd acknowledges some of this—in particular, the frame problem—in passing. He notes, for instance, that "when I see you making an arrow, a lot of things are going on. You are sitting cross legged, sharpening the arrowhead, choosing feathers to attach to the shaft, and reciting a chant. It is hard to know which parts of your behavior are essential and which are irrelevant." Sorting this out takes big brains.

There are at least two ways to see just how much brainpower is needed here. The best known involves artificial intelligence. As workers in that field have long recognized, coding the knowledge required to order lunch in a restaurant, make a fishing net, and so on, is notoriously difficult. Human beings bring vast stores of tacit knowledge to such tasks, including knowledge of which aspects of any situation are relevant to the situation at hand and which are beside the point, and the extent of these stores becomes apparent the moment we attempt to encode humanlike behavior in a computer program.

The second way to see just how cognitively challenging high-fidelity imitation is involves an experiment. (For present purposes, the experiment can be considered a thought experiment, but see the reference below for a review of the empirical literature.) Imagine that through intensive interaction and instruction, we try to slowly teach an individual from another species some sophisticated task. We then ask whether other individuals of the species can successfully imitate the behaviors of the tutored individual (the model) that are required to perform the task. This experiment might fail for two different reasons. On the one hand, the model may not be sufficiently smart to learn the task despite our patient instruction. On the other hand, the model may learn to perform the task but the other individuals may not be sufficiently smart to imitate the required behaviors. They may be capable of only low-fidelity imitation or none at all. (Or they may simply acquire the required behavior at the same slow rate as individuals who have not been exposed to the model.) For many tasks in many species, it seems likely that the second factor may play a part. (See Subiaul 2007 for a review of the complex history of these and similar experiments in monkeys.)

The possibility that these two ideas, big brains and high-fidelity imitation, might not be so distinct further implies that one may be able to blur more formally the evolutionary-psychological view and Boyd's view. A simple (and thoroughly hypothetical) model captures this claim. Imagine that human beings evolve by natural selection to be reasonably able imitators of the behavior of others. Now this might create stronger selection pressure on potential innovators to be even more skillful innovators. That is, natural selection might now drive people even further into the cognitive niche because it may now pay for innovators to innovate in ways that are more difficult for freeloaders, especially unrelated freeloaders, to copy. (This could explain why innovators are willing to bear the cost of innovation. The products of their efforts are *not* freely available to everyone.) But these enhanced innovation skills would in turn select for even-better imitation skills, and so on. In the language of evolutionary psychology, one can imagine a virtuous cycle in which an ever-more-sophisticated innovation module drives the Darwinian evolution of an ever-more-sophisticated imitation module, and vice versa. Now I am making up this model wholesale, and I don't take it particularly seriously. (And for all I know, it already resides in the literature.) But I am trying to make a larger point with it: there are ways of blurring the evolutionary-psychological view with Boyd's view.

HAYEK AND RESPECT FOR TRADITION

My second concern is one that is perhaps unexpected from an evolutionary biologist. One of the most interesting empirical findings that Boyd discusses involves the inability of many people in many cultures to explain rationally why they obey certain taboos. He considers several cases at some length, including, for example, Fijian women who refrain from eating certain types of seafood. To summarize, when Fijian women are pregnant or lactating they typically avoid, by tradition, eating sea turtles and other marine species that are consumed regularly by other members of society. When queried about *why* they avoid these foods, Fijian women often offer fantastic explanations of what might befall them or their babies if they did consume them. (For example, their babies might be born with rough skin.) Despite these nonsensical explanations, the taboo proves to be rational after all. Women are correctly avoiding the consumption of foods that contain a high concentration of a specific toxin: the ciguatera toxin, which accumulates in long-lived marine species like sea turtles and in species near the top of the food chain. As Boyd notes, consumption of this toxin by human beings can cause vomiting, diarrhea, and joint pain, among other symptoms. (The toxin appears to be especially serious for fetuses and infants.) Surveying several such taboos, Boyd concludes that norms and taboos may often be rational although they cannot be justified sensibly by any member of the community.

The point is reminiscent of one made by the economist Friedrich Hayek. Hayek famously argued that social norms, ethical mores, and even institutions are sometimes a product of a sort of social evolution, a Darwinian natural selection that acts among groups that embrace different norms. Some of these norms yield groups that flourish while others yield groups that fail. Present norms, mores, and institutions are not, therefore, a random sample of all possible ones; instead they represent a thoroughly nonrandom subset that yields reasonably functional societies. Hayek also argued that this species of social evolution was seen, at least loosely, long before the publication of *The Origin of Species* by thinkers he called "Darwinians before Darwin" (Hayek 1984a). These thinkers, he said, included Bernard Mandeville and David Hume (and the Scottish philosophers generally), as well as the Irishman Edmund Burke. (Burke's 1790 book *Reflections of the Revolution in France* is an extended meditation on the consequences of suddenly jettisoning those norms and institutions that grew up gradually through social evolution.) Some Darwin scholars, including Secord (2008), have argued that Darwin was influenced by his reading of such political theorists and economists.

Hayek further argued that in a world characterized by both tacit and dispersed knowledge, individuals often cannot justify rationally why traditions take the form they do. Indeed, in some extreme cases such justification might even be beyond the cognitive powers of individuals. A classic example involves markets and the price mechanism. No single individual can explain why the price of any product in a market, for example a new Jeep Cherokee, has the value it does. The reason is that the Jeep's price is affected by a bafflingly large number of factors that interact in often inscrutable ways, for example the supply of steel, labor unrest in rubber-producing countries, the rise of Uber, the demand for plastics in the computer industry, the performance of the S&P 500 last year, and so on. Nonetheless, the market weighs these factors, seemingly magically, and yields a market-clearing price for Jeeps. This social institution, the market, thus arrives at a solution that is unavailable to any individual reasoner or, indeed, to any central committee of planners. (Hayek and other economists of the Austrian school often referred to the [futile] task of setting prices for products via deliberate ratiocination as the socialist calculation problem. The inevitable consequence of such "rational" mispricing is shortages or surpluses of products.)

To Hayek, then, there should be a "presumption in favor" of norms and social traditions. Whether or not we can justify them rationally, these traditions and norms are the time-tested results of social evolution. They work, at least often enough and well enough to have allowed our survival as a society to the present day and to have outcompeted the many alternatives attempted by other groups through time.

There is a further point that is less well known. Hayek also believed that a legitimate science of humanity must be grounded in the recognition of that undesigned knowledge that is often embedded in tradition. In one of his excursions into the philosophy of science, especially his 1952 book *The Counter-Revolution of Science*, as well as in a number of essays collected in *The Essence of Hayek* (Hayek 1984b), he argued that while the natural sciences demonstrate the power of reason, the social sciences must be grounded in a humble acknowledgment of the *limitations* of individual reason: traditional ways of doing things might well be best for reasons that no individual comprehends. We must therefore overcome what he called the "pretence of knowledge," the hubristic assumption that individual reasoners can design solutions that are superior to the undesigned ones arrived at by cultural evolution.

Now Boyd knows all of this. Indeed, he briefly considers Hayek in his 1993 paper with Peter Richerson, "Rationality, Imitation, and Tradition." My claim is certainly not that Boyd's and Hayek's views are identical. They are not. Boyd's view is mostly descriptive: people are powerful imitators. And Hayek's view is at least partly normative: people *should* be good imitators. That is, people should be respectful of traditional ways of doing things, whether or not they understand why. Moreover, the precise forms of cultural evolution involved in Boyd's and Hayek's theses likely differ. (Unlike Boyd, Hayek is often frustratingly unclear about where "advantageous mutations," that is, successful innovations, come from. Was some individual initially an exceptionally good reasoner, or do societies somehow engage in random experiments?) Finally, Hayek's views are presented in prose that is admittedly sometimes vague (though typically stylish), while Boyd's views emerge from often-dense mathematical models (as one certainly sees in his 1993 paper with Richerson).

But I do suspect that there is a connection between these two views, arrived at in such different ways. And I do wonder if we scientists tend to think of Hayek less often than we should. (We evolutionary biologists may be especially guilty of this; see Orr 2009 for more on this.) Moreover, the reasons that we may think of Hayek less often than we should may be mostly orthogonal to the problems at hand. In particular, Hayek may be overlooked in part because he is out of fashion politically, especially in academic circles. There is, after all, a perception that Hayek used his views to justify a political order that revered tradition more than many of us are comfortable with. (Though it is worth remembering that Hayek also penned the essay "Why I Am Not a Conservative." We should be wary of simplistic stereotypes here.)

In the end, we may be left in an odd position in which we have come full circle without realizing it. If, as Hayek believed, evolutionary thinking arose first in the social sciences during the course of speculation about cultural change through human prehistory and history, are we now merely reintroducing evolutionary thinking onto its native soil? By building explicit evolutionary models of change in human culture, are we merely re-Darwinizing those "Darwinians before Darwin" that Hayek spoke of?

CHAPTER 4

ADAPTION WITHOUT INSIGHT?

Kim Sterelny

I agree, and have long agreed, with the main themes of these lectures. In particular, I accept four core proposals of the program Rob has articulated.

1. Humans (and the more recent hominins) are indeed outliers in just the ways he describes. Moreover, by the standards of evolutionary change, our lineage has become an outlier rapidly and recently. It is not just the magnitude of the difference between humans and our relatives that needs explanation. The pace of change is also remarkable.

2. Humans are both numerous and cosmopolitan be- cause we can adapt to exploit the resources, and mitigate the dangers, of a staggering range of envi- ronments. This local adaptation is primarily informa- tional and behavioral rather than physiological, and it is mediated by cultural learning. Thus the evolution of human social learning, and especially cumulative human social learning, is one key factor in explaining the difference between humans and our relatives and ancestors. This local adaptation depends on multi- generational, population-level processes. Humans are indeed exceptional in our individual adaptability, in our ability to represent and understand our envi- ronment, and in our ability to plan—to represent and respond to future contingencies. But on their own, these cognitive capacities do not suffice to explain

the size of human populations, our ecological and geographical spread, and our social and technical adaptations to our specific environments. They may be necessary (I think they are; Rob hedges his bets), but they certainly are not sufficient.

3. Relevant local information is not enough. The size, range, and ecological versatility of human social life depends as well on cooperation, which is rare in mammals but extraordinarily richly developed in our lineage. Humans are obligate cooperators, and Rob is right in thinking that human cooperation requires special explanation. Even small-scale human societies are not, and probably never have been, composed mostly of close kin. So evolutionary models of cooperation that explain cooperation between close relatives (as in the social insects) are not plausible explanations of human cooperation. Nor is direct reciprocation plausible as a general account. Withdrawing one's own cooperation in response to cooperation failure is too blunt an instrument to sustain cooperation, once cooperative interactions involve more than three or four individuals. Human cooperation requires special explanation.

4. In addition to these substantive theses, Rob and Peter have identified the right methodology for investigating the emergence of human social life: we advance our understanding through a combination of the comparative analysis of ethnographic and archaeological data; experimental work on individual motivation and cognition; and links between these data streams and formal modeling. These models help us understand how patterns in individual decision and action scale up at the level of populations, and how such populations then change over time.

So, no damning critique from me. My commentary will be on the nuances of the picture and on the further questions to which it gives rise. It will be organized around three issues: the role of individual intelligence in social learning, interpreting the models, and extending the picture across time and across different social learning domains.

1. COLLECTIVE INTELLIGENCE WITHOUT INDIVIDUAL INTELLIGENCE?

I agreed above that our distinctive human intelligence was not *sufficient* to explain our ecological flexibility and adaptability. There are hints that Rob suspects it is not even necessary, at least when we focus on the transmission (as distinct from the origins) of cultural variants.[1] For in asking for an explanation of the power of cumulative social learning to build adaptive responses to the environment, he writes:

> But I think the common understanding goes like this. Innovation is hard. It not easy to learn that nardoo is poisonous, and even harder to figure out how to detoxify it. Determining whether behaviors are beneficial is relatively easy. So once an innovation occurs, people understand why it is beneficial, and it spreads.

Boyd calls this the "library model," and he soon makes it clear that he disputes this common understanding, using Henrich's poster example of a Fijian food taboo (Henrich and Henrich 2010), backed up with a further example about hut design. That taboo prohibits pregnant women from eating marine animals high in the food chain, thus protecting them from ciguatera poisoning. The taboo is moralized; it is not just prudent advice. Henrich's ethnographic evidence suggests that the practice is stable despite the agents' having no clear idea why the taboo is adaptive. The practice seems to be transmitted from parents and elders with vague warnings about the dangers of its violation, rather than, say, from the testimony of those who have violated it and suffered as a consequence. The lesson is that an adaptive practice can be stabilized by cultural learning (and perhaps can even originate) without anyone having much of an idea why it is adaptive, not even in crude terms of probable consequences. Marvin Harris built a research program explaining taboos and food preferences (especially unexpected ones) as tacit adaptations (see, for example, Harris 1985). But Henrich and Boyd have an explicit model of such tacit adaptation.

Boyd sees agents as being strategic in their choice of models, and in the degree of trust they place in those models. So social learning is informed social learning, but most of the intelligence is in the choice of model, rather than in understanding the information that flows from the model (such understanding might be useful in reducing noise, but it is inessential). I have more time for the library model. For I put some emphasis on the fact that first, understanding why a practice is adaptive is a gradient phenomenon, and second, understanding is often partially tacit; it is know-how rather than explicit declarative knowledge. Boyd's local Fijian builders could not verbally explain why they thought that nonstandard hut designs would fail in a cyclone. That does not mean that they would have no idea, in practice, how to strengthen them. Of course, their lack of explicit knowledge is certainly not evidence that they had implicit know-how. The point, though, is that questions and surveys do not tap into know-how. So there is a lot of territory between an explicit, fully articulated understanding of the causal basis of a practice, and blind, trusting imitation. Very likely, actual cases of social transmission vary in the extent to which social pickup is accompanied by, and depends on, intelligent assessment of a practice. But my hunch is that it plays a critical role in a central range of cases: the social transmission of skill.

Boyd reconstructs the intelligent-understanding model of social transmission, in the case of the food taboo, as predicting a social dynamic something like this:

> On the "culture-is-a-library" view, the answer should go something like this: the person they learned from gave them good reason to believe that pregnant and

lactating women shouldn't eat moray eels, turtles, and the like, and they then acted on those beliefs. This belief needs to be sufficiently compelling that a woman who normally savors the rich taste of turtle will nonetheless stick to boiled cassava when she is pregnant because she is concerned about the effects on her unborn child. And it must be sufficiently compelling that she would scold a pregnant woman who is eating moray eel, even though this may offend the pregnant woman and her kin. These beliefs about the causal relationship between food and illness would be maintained because the world provides evidence that they are true. For example, somebody in the village might notice that a woman who gorged on turtle before she knew she was pregnant subsequently gave birth to a child with developmental problems.

His ethnographic data indicate that nothing like this happened. The transmission of this norm depended on a much simpler and more trusting social learning heuristic.

How typical is this example? Consider instead skill transmission by demonstration and practice. Demonstrating a skill (as distinct from merely manifesting that skill under observation) requires agents to represent the causal structure of their practice. I am no knapper, but I have had a couple of lessons. My teacher understood the structure of his own skill, and he needed to in order to demonstrate and advise. He needed to give advice on striking angles, choice of raw materials, the geometry of the rock, and the reduction sequence at which to aim. Peter Hiscock has argued, to my mind very convincingly, that active teaching of lithic skills is probably very ancient, because unguided trial-and-error learning is so very dangerous. Chips are very sharp, and a misaligned strike can result in their flying off the core fast and unpredictably (Hiscock 2014). Likewise, Louis Liebenberg has documented the cognitive requirements of the highly skilled trackway reading of southern African foragers. Their tracking synthesizes very rich pattern-recognition skills and very

careful observation with very rich natural history information (Liebenberg 1990, 2008, 2013; see also Shaw-Williams
2014). The trackers understand why the tracks of one
and the same animal change as the animal becomes tired,
alarmed, stressed, or relaxed. The social transmission of
these skills probably depends on intelligent apprenticeship:

> Adolescents and young men watch the best hunters
> in their group discuss and interpret spoor. From the
> spoor, skilled trackers can deduce an individual's age,
> sex, physical condition, fatigue level. . . . Such feats are
> accomplished, in part, by knowledge of the particular
> species' habits, feeding preferences, social organisation
> and daily patterns. (Henrich 2016, 76)

Indeed, the whole economy of foragers requires them to
have very rich natural history and local geography data files
for their territories. Boyd notes this himself in recounting
the sad but instructive story of the Burke and Wills expedition: as he notes, a "Yandruwandha 'Natural History Handbook' would have run to hundreds of pages with sections on
the habits of game, efficient hunting techniques, how to find
water, how to process toxic ferns, yams, and cycads, and so
on." They do not learn all this anew with each generation:
social learning is essential. But in their practice as foragers,
they are continually testing and refining this information.
Technique, technology, and local knowledge are repeatedly
checked against the world. There are relatively direct means
of correcting errors, for these agents receive a signal from
the world if their information is out of date or has been
distorted in transmission. A region's natural history and
microgeography are not static: creeks change their course;
water holes silt up and new ones form; animals change in
their habits and abundance. Foragers learn about all this
from their peers and elders, but they also learn from the
world itself (Hewlett et al. 2011). Rules—like norms and
taboos—can be accepted and acted on in blind faith. If a
taboo guards women against a danger that is past or was

never real, her error signal will be at best subtle and ambiguous. That is not true of the skills and information that guide intervention in the physical and biological world (this is a main theme of Sterelny 2012).

In my view, then, the food taboo contrasts with an important range of cases in two ways. First, it is an example of pure social transmission. With the exception of a few informants who reported on personal experience, the vast majority of those who conformed to the taboo received all their information about the danger of high-food-chain fish from social sources. In contrast, skill transmission is hybrid learning. Social input is essential. But no one learns how to track or how to make a hand axe just from being told, or even shown, how to do it. Skills are acquired by socially supported interaction with the world, and agents learn by doing, but in ways shaped by their social environment. Information flows from the material environment, but with the aid of models, demonstrations, advice, peer support, and examples of success and failure—for instance, a technical vocabulary that makes salient subtle differences in raw materials (Stout 2002). In this domain, social and individual information must be dynamically synthesized.

Second, while the Fijian taboo is a good example of cultural adaptation, it is not an example of *cumulative cultural adaptation*: an adaptive capacity, like the control of fire, beyond the range of individual learning, built by incremental improvement, probably over many generations. The power and versatility of local human social adaptation is largely explained by *cumulatively built* responses to various human environments, and the taboo does not exemplify this. Nor do Boyd's examples of animal architecture show that simple social learning rules can build complex adaptive capacities. These are not examples of learning-mediated adaptation at all. They are examples of adaptive action without causal understanding; they are not examples of learning, still less high-fidelity social learning, without causal understanding.

As I noted earlier, understanding why a particular practice is effective is a matter of degree. Many expert knappers have very little grip on the material sciences that explain why some forms of rock, but not others, can be shaped into useful tools. They have robust dispositional knowledge of the substrates on which they work (if I strike here, at an acute angle, such-and-such will happen), but over a limited range of forces and circumstances. The same, I conjecture, is true of the tracking skills Liebenberg documents. Those African trackers understand the processes through which animals leave traces in substrates, but only over a limited range of animals, substrates, and circumstances. Their expertise, for example, might not translate to tracking in the snow, though of course high-latitude foragers certainly had such skills. Even the Fijians are not wholly acting on blind faith. As Boyd reports, they believe that taboo violations are likely to result in sickness, and they have a general understanding that eating fish can result in illness and that some fish are more dangerous than others. This more general knowledge adds plausibility to the taboo. It fits in with their general, experiential conception of how the world works.

The library model is not right. Social learning strategies that assess the agent, rather than the cultural variants the agent manifests, can and do support adaptive responses to local circumstances. But a program that models agents as having a reflex-like urge to blindly copy their neighbors (though perhaps well chosen as models), an urge trumped only by very clear signals from the world, seems wrong too. In particular, that program downplays the role of hybrid learning, of agents not choosing between social and physical signals but combining them. In my view, hybrid learning is especially important in skill transmission, so in considering the cognitive demands on the social transmission of cultural variants, we should explicitly articulate the ways those demands vary in relation to the nature and complexity of the cultural variant.

2. MODELS AND THEIR INTERPRETATION

Rob Boyd supports his ethnographic examples with a simple formal model of the evolutionary dynamics of social learning. He calls the section in which he describes this model "The Evolution of Blind Imitation" and describes it as vindicating the idea that "selection can favor a psychology that causes most people to adopt beliefs *just because* others hold those beliefs"; it can favor a psychology in which agents are "intrinsically motivated to adopt the beliefs of others." But it is not obvious that we need to read the model's results this way. The model explores the dynamics of social learning in a noisy world. Agents' payoffs depend on making the right choice in specific states of the world. They have direct but noisy information about that state, but they have unambiguous information about the choices other agents have made (agents who, likewise, have noisy information about the world). Unsurprisingly, the noisier the direct signal—the more difficult it is to learn about the world directly—the more agents rely on an indirect, social signal.

To me, this is a simple model of intelligent strategic learning. Agents self-assess. If they take themselves to have reliable direct signals of the state of the world, they use these signals. If not, they use the information-pooling advantage of observing a population of their fellows choosing a response.[2] That is their best option. But when agents are acting optimally in choosing between individual and social learning, in order to explain their behavior we do not have to assume any intrinsic motivation to adopt the beliefs of others. Rather, within the simplifying limits of the model, these agents are maximizing their chances of getting it right. To show an intrinsic motivation to adopt the beliefs of others, we would need ethnographic evidence of agents adopting locally prevalent beliefs, even when they had more reliable evidence available that supported a different assessment. Moreover, we would have to take into account both the costs of that information and the reputational costs of being a local maverick.

Religion, obviously, seems to provide many examples of agents ignoring signals from the world in favor of social signals. But it is also a domain in which the costs of being a maverick may well be very high. That said, I would be surprised to find a strong default in favor of social information. For I think Olivier Morin is right in pointing out that much social information is about ephemera: who saw what where. The information that is distributed widely and deeply over social networks is a small and nonrandom sample of the information in the public domain; the information other agents' sayings and doings expose us to. We need to filter out a huge amount of irrelevant noise (Morin 2016).

Boyd takes his data and models to support the following picture: social learning is largely the operation of a simple, conformist heuristic, one that works well in many circumstances. For agents are rarely well placed to directly sample the environment reliably and at low cost. As a consequence, we have evolved learning dispositions that make a good trade-off between the costs and benefits of social learning. We have a bias in favor of fairly simple social learning rules, but when individuals from time to time find themselves in informationally privileged circumstances, they can, often enough, recognize their fortunate informational circumstances and override the bias. Olivier Morin's *How Traditions Live and Die* argues vehemently against this perspective on social learning. Morin is utterly unpersuaded, contending that social learning is important but strategic, depending on agents' assessment of both their own informational circumstances and those of potential models. We neither should nor do use simple conformist heuristics (Morin 2016).

Insofar as Morin's complaint is about the simplicity of Boyd's picture, it is misplaced. Real agents are more complex than Boyd's modeled agents. Of course: one aim of a formal model is to strip away inessential complexities. So the crucial issue is, would those differences make a difference to the population-level dynamics? There are at least three options we need to explore, and these can be explored only through formal models that represent the difference between simple-default social learners and Morin's highly strategic social learners. One possibility is that it does not make much difference. Perhaps populations of trusting social learners and populations of discriminating social learners would have broadly similar dynamics: similar at the level of resolution offered by comparative ethnographic and archaeological data. A second possibility is that cumulative cultural learning might be rare (or even nonexistent) in a population of strategic, discriminating cultural learners. Agents' confidence in their own judgment might make it more difficult to preserve current levels of cognitive capital reliably enough for the next generation to build on. A high level of social trust might be necessary for cumulative social learning, and the ethnographic and archaeological evidence of cumulative adaptation would then be evidence for trusting social learning in traditional social worlds. A third possibility is that modeling might show that discriminating social learning is a *more powerful* engine of cumulative adaptation than trusting social learning. Such a population might generate a richer stock of microinnovations, identify and take up innovations more reliably, and weed out errors more efficiently. Such models might show that such a population is less apt to be stuck on local optima, to be trapped in inefficient Nash equilibria, or to fail in response to environmental change. There would then be a genuine and very interesting empirical difference between Boyd's picture and that of Morin. Does the archaeological and ethnographic record show widespread (though perhaps domain- or environment-specific) adaptive failures of the kind we would expect if human populations were populations of trusting social learners? Or does it show quite reliable escape from local optima? Without formal modeling, we cannot tell whether there is a difference in their views that has empirical consequences. We see here the importance of Boyd and Richerson's methodological program: the combination of verbal analysis, ethnographic and experimental data, and formal models. Of course, these do not exhaust the options. A fourth possibility has been suggested to me by Ron Planer and Dan Dennett: perhaps social learning heuristics were initially simple but have become more nuanced over evolutionary time.

3. EXTENDING THE PICTURE

One issue that Rob explores only a little in his lectures is the extent to which we should expect social learning strategies to vary across different domains and across time. Let me begin with a few remarks on time. One of the striking features of the hominin archaeological record is that it seems to show a very long period of an extremely slow accretion of technical skills. From about 3.5 million years ago to around four hundred thousand years ago, the record shows just the addition of hand axes and fire to the original, Oldowan lithic technology. Between around four hundred thousand and one hundred thousand years ago, the record shows the introduction of new stone-working methods and of composite tools, and after one hundred thousand years ago, we see accelerating change in technology and technique. Until fairly recently, these apparent differences in the capacity to make, keep, and accumulate cognitive capital were explained by appeal to genetically based differences in individual cognitive capacity. After the publication of a landmark paper by Sally McBrearty and Allison Brooks in 2000, this form of explanation has been supplemented or supplanted by explanations that appeal to differences in social organization and social network size (Henrich 2004; Powell et al. 2009). Richerson and Boyd have been part of this shift (Richerson and Boyd 2013). One option they have not yet systematically explored, but which fits very well into their overall program, is that we see in this record the cultural evolution of improved techniques and tools for social learning. We have been *learning socially* how to improve the ways we learn socially. Celia Heyes has pressed this idea particularly hard (Heyes 2012).

So the character and reliability of social learning are not constant over time. The same is true across domains. It should be no surprise that cumulative cultural evolution by and large failed to develop effective folk medicines (despite a few famous exceptions). Much traditional medicine is terrifying; for many splendidly appalling examples, see Edgerton (1992). The pathogenetic environment is very labile, changing not just because of the rapid evolutionary response of pathogens, but because the identity and virulence of pathogens change in response to changes in population size, economic activity, and movement patterns. Most obviously, a change from mobile to sedentary life exposes agents to new pathogens, as people live close to their detritus and excrement. But so do changes in population size and connectedness. These determine whether pathogens spread to new hosts more rapidly than immune mechanisms and/or the death of previous hosts cause their local extinction (see Ewald 1994). The construction of malaria-friendly environmental conditions through changes in agricultural practices is one of the poster examples of niche construction discussed in Odling-Smee et al. (2003). Moreover, the medical environment is not just fast changing; it is complex, with long causal latencies. Many agents recover from illnesses despite interventions that do them no good (and may even do harm); others die even though an intervention improved their chances. Premodern society saw a much more reliable accumulation of weapons expertise than medical expertise. I doubt whether that was because we were under stronger selection to kill than to cure. Rather, with weapons technology, successful innovation and the recognition of success are much more tractable. The success or failure signal of a technical innovation is often quite immediate. The edge cuts with less effort; it keeps its edge longer; the binding holds a point firmly in place. Or not. Moreover, technologies in regular use will generate a lot of trials. If there is variation in the local group (as there will be, if there is an innovation), there will be some comparative data.[3] In general, innovations in technology and technique involve relatively short, relatively simple causal chains that link the use of a tool to an outcome. Furthermore, agents often have some information about a significant number of trials. These are only rough and ready generalizations, as Boyd's example of Fijian house design shows. Those builders find out rapidly whether their house is stable and weatherproof. But it might take decades to learn whether it is storm proof.

The assessment of norms and normative innovations, I suspect, is somewhat intermediate between that of medical expertise and hunting technologies. Even if we assume that agents understand and value the social peace and mutually profitable cooperation that norms can make possible, it is typically hard to assess the social impact of one norm in comparison to another. Consider Alvard's classic example of the elaborate norms governing the division of a whale after a successful whale hunt (Alvard and Nolin 2002). The participants probably did not have much opportunity to see the effects of alternative norms on social peace, on cooperation, and on investment in whale-hunting equipment and techniques. Moreover, a change in the norms of division might well manifest its effects slowly, even over a generation or more, and in conjunction with other changes. Members of the community very likely have experienced disputes and discord when the norms are broken or stretched, so they might well understand the costs of a collapse of the normative regulation of the division of the proceeds. But they are much less likely to be able to assess the results of normative tinkering.

Perhaps for this reason, Boyd and his colleagues tend to model normative evolution as an arbitrary within-group process. To the extent that we see adaptive normative packages, that is the result of selection between groups. I agree that human cultural groups probably do form a Darwinian population and evolve by cultural group selection. But (1) these metapopulations are small; (2) generations at the group level turn over fairly slowly; and (3) if norms tend to form complexes or systems, variation will be relatively constrained. So we would not expect this process to be very efficient. On these suppositions, cultural group selection will be a fairly blunt instrument, and so normative packages are unlikely to be very efficient. If normative packages are more efficient at sustaining a cooperative social life than we would expect from a noisy, inefficient selection machine (though how one might show this, I have no idea), perhaps this might be explained by a prosocial bias in the generation of norms. Boyd and Richerson's models and human history both show that some astonishingly destructive norms can be stabilized by punishment and internalization. Even so, it has been argued with real plausibility that it is easier to teach and to stabilize norms that are in tune with our social emotions of sympathy, empathy, and fairness (Nichols 2004). If so, the cooperation-promoting character of many normative packages might be the result of two, perhaps weak, but mutually reinforcing biases. Cooperative packages are somewhat more likely to establish in a local group, and such groups are somewhat more likely to survive and spread.

A final thought on norms. We can think of norms and norm psychology in two ways. One is exemplified by Alvard and Nolin's account of the norms of whale division. These norms of division are quite complex, explicitly articulated, known to the participants, taught explicitly, and a matter of local lore. Penalties for norm violation, likewise, are explicit, public, taught, and often quite precise and are themselves part of the normative package of the group. Alternatively, we can think of norms as much less cognitively sophisticated. Resentment norms (as we might call them) are expectations of default patterns of action (what we expect a group mate to do, say, if the fire is going out), coupled with a disposition to respond with anger, disapproval, hostility—in short, with resentment—if that default expectation is not met. Boyd's examples of norms, and his analysis of the ways in which norms can sustain cooperation even in societies without institutions, is premised on norms in the rich sense.

The Turkana norms are explicit. Yet clearly norms in this sense cannot be the foundation of the distinctive forms of human cooperation. Language, for example, is a product of characteristically human cooperation, not its precondition. What remains to be shown, then, is that the Boyd-Richerson analysis of the role of norms in human cooperation can be redone premised on resentment norms; on a psychology of interaction that might plausibly be in place before the emergence of distinctively human cooperation. Consider, for example, Boyd's solution to the higher-order free-rider problem: punishing those who fail to punish first-order norm violations. He points out that the appropriate punishment of norm violation is itself part of the normative system. Whatever psychological and social mechanism supports first-order respect for norms, and punishment for norm violation, also supports respect for norms of punishment, and punishment of normative failures to do with obligations to punish appropriately. There is no distinct phenomenon, requiring its own explanation, of second-order free riding or second-order punishment. While this is true of a fully articulated system of norms,[4] such as those of the Turkana, it is by no means obviously true of resentment norms. There is no automatic mechanism that ensures that those agents resenting expectation failures also resent those who fail to respond with hostility to others' expectation failures. In contrast to the Turkana, at this stage there is no explicit, articulated system that agents buy into as a whole. Default patterns, and expectations based on those default patterns, develop piecemeal.

4. OVERVIEW

These questions are not meant as a challenge to the core ideas that Rob articulates in these lectures, or to the program that those lectures represent and sample. As I wrote at the beginning of this response, I agree with the most central elements in that program. To some small degree, this accord is the result of convergent thinking. But it is mostly testament to the extent to which Rob Boyd and Pete Richerson have shaped[5] everyone's views of the evolution of the distinctive features of human social life, and of the capacities that support that life. In raising the questions above, I take myself to have helped show that their program is stimulating, productive, and tractable. It does not just suggest important questions; it opens up ways of answering them.[6]

CHAPTER 5

INFERENCE AND HYPOTHESIS TESTING IN CULTURAL EVOLUTION

Ruth Mace

I am delighted to take part as a respondent to these lectures. Rob's 1985 book *Culture and the Evolutionary Process*, written with Pete Richerson, was a formative book in my academic life. The first book that really changed what I decided to study was Richard Dawkins's *The Selfish Gene* (1976), which I read as a teenager. That book inspired me to think about social evolution and to go and study Zoology at Oxford, which was then, and still is, one of the head-quarters of adaptationist, inclusive fitness thinking. Many years later I read *Culture and the Evolutionary Process*. I am not particularly mathematical, so I skipped most of the equations, despite those being about half the book, but the general ideas were very thought provoking. By then I had moved to London and become an anthropologist. At that time, the field of human evolutionary behavioral studies was generally considered to be divided into three schools of thought: evolutionary psychology, human behavioral ecology, and cultural evolution. Affiliation with each of these three groupings was generally predicted by which disciplines people came from, which topics they were interested in, and which methods they used; but of late these distinctions have blurred. I considered myself a human behavioral ecologist.

Behavioral ecology is an absorbent discipline with a strong tradition of empirical testing of evolutionary hypotheses, and it drags in approaches and methodologies from any scientific field that helps with that task. Unlike evolutionary psychologists, human behavioral ecologists never made claims about human universals, as human behavioral ecology (HBE) is the study of human behavioral variation. They were not precise about the genetic or cultural origins of behavior, but more concerned about adaptation. Cultural evolutionary studies used to focus mostly on modeling mechanisms of learning and transmission and the implications for cultural evolution, while HBE researchers were looking at correlates of patterns of behavior in the real world. The cultural evolutionists talked about culture, and HBE researchers talked about ecology. It is not really possible to draw a distinction between ecology and culture as determinants of behavior, because culture is one of the ways that people learn about their ecology and what works best in each environment, as Rob eloquently explains in his lectures. As questions about the origin of behavior have become more specific, the agendas of the two fields have merged somewhat, with researchers attempting to understand what are the evolutionary processes that generate specific aspects of human behavioral variation, using a range of methods, and taking both cultural transmission and ecological variation more seriously. So nowadays I find it increasingly hard to distinguish between these two fields (Mace 2014).

Therefore, despite Rob and I coming at the task of understanding the evolution of behavior from different traditions, I agree with almost everything Rob says in his lectures. There is little doubt that the role of norms in guiding human behavior is overwhelmingly important. It hit me hard during my first experience of field anthropology in a small-scale society. I worked on the reproductive decisions of nomadic Gabbra camel pastoralists in northern Kenya. As a visitor, it was appropriate for me to sleep outside the nomadic huts under the stars on a mat or skin, as did all the unmarried people. That was definitely the best place to sleep given the heat and the fantastic north Kenyan night sky. But I quickly realized that it was important to sleep on my side, legs together and tucked up, wrapped of course in a full-length skirt. It was no problem falling asleep in that position, but I was always concerned that by morning I would wake up to find myself having rolled onto my back or front in what would have been a totally unacceptable position for a woman to sleep in. This was just one in a long list of norms that I suffered constant anxiety about violating. We are so familiar with our own culture that we don't always understand the difficulties others may have adjusting to it. Once when my research group in London went out for drinks after a research seminar, the conversation happened to turn to Bob Marley. It doesn't matter who Bob Marley was, but if you are English or American or European or Jamaican or Ethiopian or indeed of many other nationalities, I am guessing there is a very good chance that you know about Bob Marley. If you are Chinese you may not have heard of him. When a newly arrived Chinese colleague asked who Bob Marley was, she was met with incredulity, albeit well meaning, from the other students. She was so embarrassed that she decided not come out to drinks after the seminar again. It was months before she told me what had happened, and I was shocked when I realized how much she had felt a social pressure that the others had not intended or even perceived. These are just the norms of socializing. Imagine the potential complexities of negotiating marriages, trades, political situations, and conflicts across cultures. We are ruled by norms.

But while it is clear that norms matter, there remain many unanswered questions about how these norms arise, how locally variable they are, how they are maintained, and how easily they change. An important reason to focus on cultural transmission mechanisms is if they generate fundamentally different outcomes from other mechanisms of behavioral development and evolution. "Cultural group selection" is one such mechanism, and it is that section of Rob's lectures that I want to comment on. This may be a relatively small part of the overall thrust of the lectures, but it is an example of cultural evolutionary theory making different predictions from other evolutionary models (group selection being thought less important in genetic systems, as the integrity of genetic groups is generally destroyed by migration). Furthermore, it is a theory that has captured the public imagination as well as generated academic debate.

Cultural group selection is a hard hypothesis to test, not least because even the concept of "cultural group selection" is sometimes used to mean different things; the group structure referred to may be genetic or cultural or require groups to be both, the groups may be whole ethnolinguistic groups or a just village or band, and relatedness and/or fitness may be genetic or cultural. Much of human behavior is too variable to be usefully thought of as a group-level trait. But anthropology is, at its heart, the study of differences between cultural groups. So if there are no consistent differences in behavior between cultural groups, then our field does not really exist. I do agree that cultural group selection could be important in the evolution of some behaviors or institutions, and competition between groups could theoretically be an important selective force. So I see the main question now as an empirical one. We need to find out which, if any, behaviors vary at the group level, and on what scale. Which traits, if any, are selected for by competition between groups? And how? Is there enough time for group selection to be significant on the timescale over which cultural change occurs?

The hypothesis that warfare promotes in-group altruism due to cultural group selection now seems to be a widely accepted theory—you can read articles in scientific and popular journals, where it is described as accepted wisdom. The idea seems to make intuitive sense. My parents used to talk about the "Dunkirk spirit," referring to the motivation of hundreds of civilians who set sail in their own small boats, at great risk to themselves, to rescue British soldiers trapped on French beaches who were retreating from the German armies invading France in World War II. My father, a boy at the time, lived on the Suffolk coast and knew some of the boat owners who made the hazardous trip, while my sister-in-law's father, old enough to be called up to fight, was on the other side of the channel on the beach, queueing up to be rescued. While being strafed by enemy fire, they were under instructions that if any man jumped the queue he would be shot by his own commanders; so it seems the British Army was relying at least as much on punishment as altruism to maintain discipline (I don't remember seeing that bit in all the Dunkirk movies). The Dunkirk rescue enabled Churchill to skillfully portray a defeat as an important propaganda coup, which lifted spirits at a desperate time and may have helped change the course of the war.

In the interest of proposing alternative hypotheses, it is possible that there are individual benefits to engaging in warfare that provide the most important reasons for taking part. Turkana warriors can certainly gain individual advantages from a decision to go on cattle raids. Those who do not go on raids do not get the spoils, and without cattle you cannot marry and reproduce, so some young men who think about it strategically are likely to conclude it is a risk worth taking. Punishment for those who have broken an implicit contract to fight and not run away is a feature of all armies, and punishment is often also a feature of the rules that prohibit leaving religious or other institutions. While punishment may be a norm that evolved by group selection, it could be coercion by leaders taking advantage of power asymmetries, in which case it may not necessarily be associated with group selection or enhanced altruism. It is plausible that punishment is group selected, but whether or not punishment shows enough variation between groups and contributes enough to the public good to be selected for by differences in group success has not yet been shown and is indeed going to be hard to show. Inference from design is usually the first, but not necessarily the strongest, test of evolutionary hypotheses.

I was interested in testing the idea that competition and conflict between groups, such as interethnic warfare, leads to parochial altruism (that is, altruism directed only within a group). My PhD student Antonio Silva tested the hypothesis in Belfast, Northern Ireland, where there have been at least three hundred years of conflict between Catholic Republicans and Protestant Unionists. The conflict was abated by the so-called Good Friday agreement in 1998, but certainly not entirely ended. There are still huge walls built by government dividing the two communities in some areas, intermarriage between Catholics and Protestants is still quite unusual, and riots still break out over sectarian incidents. In many areas streets are marked with sectarian murals, or claimed in loyalist marches designed to be inspirational and/or provocative. Education is segregated, as virtually all schools in Northern Ireland are either Protestant or Catholic. There seems still to be, to my outsider's eye, some kind of collective madness governing the relationship between groups that needs to be explained.

We examined this by doing an experiment with people who helped us with a questionnaire (about their background, religiosity, and experience of sectarian threat), after which we gave them £5 and said they could donate any part of it to the local school (either Protestant or Catholic) or to a neutral charity (Silva and Mace 2014). We tried this in twenty-one areas of Belfast, some of which had high rates of conflict and some low. We also dropped "lost letters" addressed to Catholic or Protestant charities in the same areas. We predicted that increased threat would reduce altruism toward the out-group but increase altruism toward the in-group, more so than in the control (a neutral charity). In both cases we found that altruism to the out-group (i.e., the group you were not a member of), be it the return of letters or school donations, went down in areas of conflict, as predicted. War makes you hate the enemy. But conflict did not increase altruism toward the in-group. Altruism to the in-group stayed pretty similar to altruism toward the control group. Just after we finished that fieldwork, serious riots broke out in Belfast (concerning which flag to hang for how many days per year on Belfast City Hall). We rushed back to repeat the same experiments in a few of the same areas and found that altruism of all sorts went down during the conflict (Silva and Mace 2015). If conflict breeds altruism, it must be a rather small effect.

Our results surprised us, as several other studies have found increased in-group altruism in economic games played with those who have faced conflict. But on closer inspection, much of the evidence is equivocal. One study found that Israelis in a retirement home were more likely to punish each other's lack of generosity in economic games held during the 2006 Israel-Palestine war than before or after; but there was no control group or out-group in the experiment (Gneezy and Fessler 2012).

Another study found that schoolchildren with more experience of war in Georgia were more generous to their schoolmates than they were to children from another "distant school" (with whom they were presumably not in conflict) (Bauer et al. 2014); but there was no effect from exposure to war on adult behaviors in that study. Overall there are actually very few studies that use real groups, donating to real institutions in real conflicts, that also examine both sides and use a control, as we attempted to do.

Numerous studies have demonstrated how cooperative behavior is highly context dependent, variable within populations, and quite sensitive to local ecology. I am not convinced that cooperation is one of those behaviors that evolved by cultural group selection. Some other behaviors or institutions may show more stable variation within and between cultures, and over evolutionary timescales, than does the level of cooperation. Phylogenetic methods, with phylogenies based on language similarity, are one way to measure the endurance of cultural norms over time, as Rob notes. We have shown that many cultural traits, including marriage and kinship norms as well as political systems, can have quite strong phylogenetic signals of "cultural history," suggesting that mechanisms do maintain the cultural differences in those traits between cultural groups. Such institutions are therefore better candidates for traits subject to group selection. Frequency-dependent selection on individuals could also be one mechanism for maintaining cultural differences (for example, frequency dependence seems to apply in the evolution of kinship systems in China) (Ji et al. 2016). Different behaviors will have different evolutionary pathways, as I hope more and more specific empirical studies will reveal. We are currently only touching the surface of a huge empirical project that is underway, but much more still needs to be done.

CHAPTER 6

ADAPTABLE, COOPERATIVE, MANIPULATIVE, AND RIVALROUS

Paul Seabright

Human beings are the most ecologically adaptable and massively cooperative species on our planet, as Robert Boyd convincingly and enjoyably reminds us in his Tanner Lectures. We are also the most spectacularly and violently competitive, and the most deviously manipulative of all species. This might seem an incoherent description, but in fact the latter qualities are deeply implicated in the former ones. It is precisely the fact of our extraordinary cooperativeness that allows us to create the massive resource gains that provoke our competitiveness and manipulativeness.

If human life were a zero-sum game, our interactions with our conspecifics would be relatively straightforward, and we would learn to avoid doing anything that might benefit them in any way, which would likely mean that interactions would be kept to a minimum. In fact, human life is a highly positive-sum game. We therefore frequently, and quite reasonably, do many things that benefit our fellow human beings, since only by doing so can we expect to realize the massive gains from cooperation. We thereby lay ourselves open all the more easily to exploitation, when the gains from cooperation are divided much more unfavorably than we expect. Revenge is famously described as a dish best eaten cold. It can also be properly savored only by enemies who were once friends.

Added to this tension at the heart of the relationship between would-be cooperators is the fact that in the kinds of social groups that human beings inhabit, the identity of one's collaborators is constantly being renegotiated. Cooperation rarely takes place just at the level of the whole group. It also takes place within subgroups and coalitions, membership of which can be extremely fluid. So at the same time as individuals are cooperating with some members of their subgroup or coalition, they are also competing to be accepted by the powerful coalitions in the group. The tension between the need to cooperate with some individuals and the need to compete against others (or even the need to cooperate with a given individual in some contexts and compete against that individual in other contexts) can provoke highly complicated motivations and behaviors. This darker dimension to what makes us human is not something Robert Boyd denies, but it doesn't figure much in the picture he presents us. In this short comment I want to say why this is an important omission.

Boyd's lectures present an excellent summary of an important program of research that he has pioneered, together with Peter Richerson and a number of their colleagues and students, including Joseph Henrich and Sarah Mathew.[1] It deserves to be much better known among scholars in the social sciences and the humanities, since it provides a picture of human nature with powerful implications for how societies should be organized. This research program accounts for the fact that human beings have become outliers in the natural world, colonizing almost all the environments on the planet that can support life, thanks to a supremely flexible collective intelligence that makes us collectively smarter than any individuals in the population. Indeed, we do not need to be as intelligent individually as we are collectively since the most efficient mechanisms of information transmission involve imitation in which we do not have to understand the precise causal consequences of every behavior pattern that we imitate.

In Boyd's account, these characteristics have not only enabled human beings to work out technological solutions to
problems of survival and mastery of our environment; they
have also enabled us to solve collective-action problems (no-
tably the provision of public goods) by coordinating on collective norms, which are enforced by the social institutions of
punishment of individual violators, punishment itself being a
norm-governed process. These norms evolve through a process of competition between groups, in which more adaptive
norms have a better chance of being selected and thus of
spreading in the population as a whole. This is a process that
Boyd and his collaborators have called "cultural group selection," most unfortunately since it has nothing to do with
group selection as this process is known to biologists.[2]

All modelers have to simplify the reality their models describe, just as maps have to be simpler than the world they
enable us to navigate. Maps can be misleading (if they leave
out important road junctions, for example), but it is a mistake to criticize a map for being simple. Boyd's simplification
is to reduce the major ecological and social challenges human
beings have had to negotiate to the problem of motivating individuals to contribute to public goods. In this view, it is clear
that a group will be better off if some public good is provided, and if the group's leaders have decided by some process that there should be an enforceable norm determining
the appropriate contribution of each individual. Individuals
have to be monitored, and if the monitoring reveals them
to be failing to contribute, they have to be punished. Sarah
Mathew's work among the Turkana shows this process at
work very convincingly for a particular (local) public good,
namely participation in military raids against rival groups.

In the rest of this commentary, I want to concentrate on
the partially misleading nature of this map of our social life
by drawing attention to some crucial features that it leaves
out. There are three in particular:

1. It is typically much harder to monitor norm compliance
   than the public goods example suggests, leading to

frequent major disagreements among members of a society as to who is a norm violator and who is not.

2. Norms have to be communicated as well as enforced. As a result, there are frequently major disagreements between members of a society as to what the appropriate norm is, either in general or (more commonly) as applied to a particular case.

3. Although there may be a common interest on the part of all members in a society in the enforcement of some norm as opposed to no norm at all, there are typically important conflicts of interest over the question of which among several candidate norms should apply. This means that information transmission is frequently manipulated by individuals to influence the norm enforcement process to their own advantage. This adds to the difficulty of norm enforcement mentioned in (1) above.

I consider these features in turn.

1. MONITORING NORM COMPLIANCE THROUGH DIRECT SANCTIONS

As Boyd notes, there is a major problem with the idea that cooperation can be enforced by reciprocity in large groups—that is, by the threat that if some individuals fail to contribute to the public good, others will refuse to do so. The problem is that mistakes can be made in detecting defectors, and each mistake may then set off a cascade of further defections. The solution, Boyd suggests, lies in "direct sanctions, which can be targeted." In this case, he writes, "errors or uncertainty do not set off a cascade of defection, and this means that cooperation can be stable in a noisy world." It is evidently some years since Boyd last tried to find out which of a group of children was responsible for some school or home disaster. The trouble with direct sanctions is that they create a strong incentive for disputes about who is really a norm violator and who is not; it is very rare for the norm enforcer to be able to observe this directly. Boyd writes in several places as though in large groups a kind of "wisdom of crowds" operates to diminish uncertainty about who is a norm violator. But he never makes this argument explicit, and it is pretty implausible. Disputes about who really was violating a norm are the staple of soap opera. They lie at the heart of human social life, and the more earnestly the norm is enforced the more intractable the factual disputes become. Think of the viciousness of the arguments about who is the true representative of the prophet and who is an apostate, arguments that fuel the civil war within Islam that is currently tearing the Middle East apart.

2. THE IMPORTANCE OF NORM COMMUNICATION AS WELL AS NORM ENFORCEMENT

Those who seek to enforce social norms need to ensure that group members know what the norms prescribe. This is true for two main reasons. First, although the domain of application of norms seems remarkably universal across human societies—all societies have norms about killing, fighting to defend the group, sexuality and marriage, and food preparation, for example—the specific content of the norms varies substantially from one society to another. There are frequently different food taboos in neighboring societies, and different rules about who can marry whom. It can be an important element of the identity of certain groups that their norms are different from those of their neighbors (think of the importance of food taboos or elements of the Sabbath observance in helping to remind Jews of their differences from the non-Jewish populations in which they live). Important and sometimes elaborate ceremonies of communication can serve to inform and remind group members of the norms they are supposed to observe, since it is usually not possible to work these out from first principles—they are not just "obvious." And of course the more elaborate the communication required, the more often mistakes will be made and people will follow the "wrong" norm.

The second reason norm communication is so important is that even when norms are universal, their application typically varies according to circumstances. Consider the norm that men should be willing to fight and risk their lives in the defense of their families and fellow group members. I know of no society that completely lacks norms of this kind. But when is the right moment to fight, and who needs to make what kind of sacrifice at what point? Once again these are typically not things that anyone can just work out for themselves—there is in most societies a structure of authority that privileges certain types of communication about the application of the norms. Military leaders are privileged sources of information about what the general norm prescribes at any particular juncture, and there are of course further norms about when and how military leaders should be obeyed. These norms are often contested, which once again underlines how difficult it is to reduce the process of norm enforcement to one in which society's leaders simply observe who is in violation and then inflict the appropriate punishments.

Tempting though it might be to think that arguments about the appropriate norms are a product purely of large and complex modern societies, there is evidence that such arguments go back a very long way, even to preliterate times. Recent phylogenetic analysis has suggested that a folk tale called "The Smith and the Devil" dates back to the Bronze Age (Graça da Silva and Tehrani 2016). What is interesting about this tale is that it pits two strong moral norms against each other. The first is the norm of truth telling, and the second is the norm of refusing to compromise with evil people. The tale is an account of how to navigate in a world in which norms appear to be in conflict. This is a theme at the heart of the Iliad and the Odyssey, of the Mahabharata and the Ramayana, and of other myths from around the world; it is certainly not a product of the modern world.

Importantly, the fact that it is often hard to know what the general norm requires an individual to do in particular circumstances means that the way is open to norm manipulation.

3. CONFLICTS OF INTEREST IN
NORM ENFORCEMENT

A famously funny comedy sketch from the *Beyond the Fringe* team[3] shows Peter Cook as an RAF commanding officer in the Second World War talking an earnestly patriotic Jonathan Miller into undertaking a suicide mission: "What we need now, Perkins, is a really futile gesture. It could change the whole tone of the war." As Perkins leaves to board his airplane he says, "Good-bye sir—or should it be *au revoir*?" to which the answer, inevitably, is "No, Perkins." The increasing use of suicide bombing in modern insurgent warfare, as well as growing evidence about the psychological pressure exerted on the perpetrators of such acts (who include children), makes this a lot less funny than it used to be.

It is rare for disagreements about what a norm requires in particular circumstances to have no distributional implications. I benefit from the interpretation that you rather than I are the suitable person to embark on the suicide mission at this point, even if I acknowledge that the norm might require either of us in principle to be willing to do so. As the pastor of the Seabright Salvation Ministry, I benefit greatly from the interpretation of the norm that requires you to donate one-tenth of your income to the organization whose finances I oversee.

All of this means that a much larger part of the communication that takes place around norms in most societies is about individuals manipulating other individuals than you would think from reading Boyd on the subject of net making or nardoo processing by Yandruwandha Aboriginals in Australia. In these examples, all parties share a common interest in faithful transmission of information—it's hard to see what an elderly Yandruwandha might conceivably have to gain from distorting the information about how to process nardoo. But common interest in faithful transmission may not be the typical case. The belief that it is typical may lead the anthropologist to underestimate the complexity of the cognitive mechanisms with which natural selection has endowed us. If you think that evolution in prehistory required us to just blindly copy what the benevolent elders of our forager groups told us to do, you will tend to think it is obvious why we are so gullible in the face of modern manipulators. If you think that manipulators were already plentiful in prehistory, and that our ancestors had to learn to deal with them, our continued vulnerability to them today requires a much more developed explanation.

To illustrate this point, I would like to consider how narratives in traditional myths and stories bear the marks of manipulation rather than of straightforward transmission of information between parties who share an interest in the truth. A number of scholars have emphasized the common narrative structure underlying many myths and stories. For example, Joseph Campbell (1949) and Christopher Booker (2006) deploy a powerful literary insight to draw attention to the common structure of many mythical and post-mythical narratives.

The psychological explanation they favor for this common structure is entirely implausible—it is basically a story of Jungian archetypes—but it is possible to combine the common structure they document with a more convincing hypothesis, consistent with the work of Boyd and his collaborators, about why these narratives have such a common structure. Boyd's approach provides an altogether better psychology than Jung's: the domain of the narratives may be universal but the content is not. Although the narratives have a similar general structure, it is the details that count. The reason for this is that they are designed not to reinforce a general message about the norm that should be followed, but rather to help the reader or listener think through the challenge in working out which norms apply in any given situation. The narratives reflect, in other words, that it is often far from obvious what norm an individual should follow—the stories help him or her (and in most traditional myths it is usually him) to know what to do.

As examples, consider the following observations, all of which constitute evidence against the view that the main purpose of narratives was simply to reinforce the dominant norms in society:

1. "Defeating the monster" stories (one of the most basic plots, according to both Campbell and Booker) usually rely on the hero's finding some way to outwit the monster, not on superior strength. The message is not that the hero has to hang in there rather than to quit (as one might think if the purpose of the narrative were to reinforce the norm of willingness to fight to defend one's group), but rather that the hero has to think hard about how best to win against an adversary that is much stronger than he is.

2. Many dramas and stories contain stern authority figures whose status the hero successfully undermines.

3. A story with a prohibition (don't open the secret door, don't eat the forbidden fruit) always sees the prohibition flouted, with dangerous but not disastrous consequences for the hero. If the purpose of the narrative were to reinforce the norm, flouting the prohibition would always lead to disaster. Instead the prohibition stories provide a lesson to the hearer or reader in how to flout a prohibition and get away with it.

4. There are almost no deluded female figures in drama or stories until very late (essentially until the nineteenth century). Delusion in narrative is a characteristic of figures who exercise power, and very few female figures do so (the exceptions are queens or goddesses in Greek myths). Once again, if narratives were reinforcing social norms, they would not be underlining the deludedness of social authority figures.

In short, narratives do not just reinforce a given social norm but provide an imaginative exercise in applying the norm in difficult circumstances. The norms in question are often framed explicitly in response to rival norms, or rival accounts of the application of a norm. Communicating norms both requires and solicits complex social intelligence, on both sides, and can be a highly manipulative exercise.

Let me summarize the message of this discussion of normative communication for the view of culture advanced in Robert Boyd's lectures. Cultural transmission is a strategic exercise in which those who transmit information seek (not always successfully) to manipulate those who receive it. Sometimes they share a common interest with the receivers—as in much transmission of forager technology. However, forager technology is not a good model for the general case of cultural transmission—normative disputes, in particular, are very different since there are more conflicts of interest and fewer instances of fully common interest. Such phenomena go back a very long way in the story of human culture. Indeed, we will not understand why culture wars are so intractable if we think of cultural manipulation as just a modern phenomenon.

Once we understand normative communication as a complex, manipulative exercise, we will be better placed to understand why individual human intelligence has become so subtle in spite of our ability to piggyback on the collective intelligence of our fellow humans. It may have been enough for us to be only moderately individually smart in order to navigate our physical environment, for culture could do much of the work. But to navigate our social environment, we cannot count so innocently on the collaboration of our fellow humans, and to survive in our social environment we have had to learn to be very smart indeed.

RESPONSE

CHAPTER 7

CULTURE, BELIEFS, AND DECISIONS

Robert Boyd

I want to thank the four commentators for their thoughtful and challenging comments on my essays. They have some serious disagreements with my work. I am mostly not convinced by their arguments and will explain why below. But I appreciate their serious engagement with the basic project: using ideas and models drawn from population biology to think about culture and cultural change. When Pete Richerson and I began our work, this was not the case. We got plenty of criticism, but very little that was actually relevant to the core of our work.

Much was of the form "this is not the way we think about culture in my discipline." Some was less polite. In the early 1980s, a well-known anthropologist was asked to comment on a paper of mine in a large public venue, and he opened his remarks by saying, "I wish I hadn't wasted the hour it took me to read this paper"; things went downhill from there. Fortunately, not everyone agreed, and the idea that cultural evolution can be modeled using ideas from population biology is no longer revolutionary. Although the four commentators have serious misgivings about some of my ideas, they accept the value of trying to understand how culture evolved. In the end, all of us will be wrong about much of what we believe about cultural evolution, but substantive debates are the means for making progress.

The comments of three of the four, Sterelny, Seabright, and Mace, also share a common complaint, namely that the picture of human action I paint doesn't give people enough credit for making smart, well-informed decisions. In this, they represent a broad swath of thinking about society. Imagine doing a principal components analysis of scholars in the social sciences. Surely the first or second component would capture the roles of individual decisions in determining social outcomes. At the left end of this axis would sit Durkheim and much of twentieth-century cultural anthropology, scholars who place the emphasis on how society creates people. At the right end of this axis would be thinkers like Hobbes, modern rational choice theorists, and behavioral ecologists, scholars who see society as the result of individual choices. The comments of Sterelny, Seabright, and Mace fall somewhere on the right side of this axis. Sterelny thinks that people are smart about what they learn from others, Seabright thinks that people are crafty manipulators of norms, and Mace thinks that cooperative actions like those of Turkana warriors can be explained in terms of individual interests.

I have lots of sympathy for right-side thinkers. People work hard to further their own interests, and theories that ignore this fact get lots of stuff wrong. The problem is that people make choices given their beliefs about the world, and right-side theory usually doesn't explain why people's beliefs about the world are correct. This is clearest in modern rational choice models like those that dominate economics. Actors have preferences, formalized as utility functions, and beliefs, formalized as Bayesian priors. At equilibrium, actors choose actions that maximize their expected utility given the choices made by other actors. Economists typically assume that people have correct priors, and so, on average, people know what they are doing and make sensible choices. Other disciplines are less explicit but share the conviction that people generally know what they are doing. The catch, I think, is that people get beliefs from their social world. When I was young I believed that you had to wait an hour after eating before going into the pool; otherwise you would get cramps and drown. I believed this because it was what everybody believed. I now believe that you can go into the pool right after eating because medical experts say it is fine, and I let my own kids risk death on my new conviction. Neither belief was based on any direct evidence. True, I've seen kids eat and then swim, but what if doing so leads to a one in one hundred chance of death? An unacceptable risk that my own experience cannot preclude. My mother firmly believed that a person's astrological sign provided useful information about that person's character. She believed this because other people around her believed it, and she also thought her own observations confirmed that the belief was correct. I think this is all nonsense, mainly because of what the people around me think. My students and I spend our time thinking about evolution and behavior; two hundred years ago we would have been puzzling about the Trinity. There's no doubt that the individual choice approach is useful once you specify what people believe. But when you step back, you need to explain why people have the beliefs they do, and this depends on the social context. My claim is that the approach in my essay explains why people are built this way.

H. ALLEN ORR

Allen Orr raises two important issues in his commentary: the relationship between cultural learning and cognitive ability, and the relationship between my work with Pete Richerson and that of the Austrian-school economist Friedrich Hayek. In my essay I did not do proper justice to the complex relationship between cultural learning and intelligence and said virtually nothing about the relationship of our work to the work of other authors. I am grateful to Orr for the opportunity to expand on these two issues.

I agree with Orr that there is every reason to believe that cognitive abilities and cultural learning are mutually reinforcing. More-powerful cognitive abilities make cultural learning more effective, and cultural learning increases the benefits associated with increased cognitive ability. Cultural learning is fundamentally an inference problem. Enhanced cognition, particularly theory of mind and causal reasoning, help people draw accurate inferences about the behavior of others or the relationships between events. Knowledge of others' minds, sometimes called theory of mind, is likely to be particularly important for teaching. Consider how a naive learner learns from a model. The model has some mental representation—the word "bird" refers to things like robins, as well as things like penguins and ostriches, but not bats. The learner observes the behavior of the model. When the model says "look at that bird on the hibiscus bush," the learner has to infer what the model has in mind.[1] People seem to have a more fully developed theory of mind than other primates. Causal reasoning, folk physics, and intuitions about materials may help people invent useful tools that can then be copied by others. It also seems likely that cultural learning allows investments in cognitive abilities to yield larger fitness increases. The ability to learn from others gives cultural learners access to more information than they could acquire on their own, and access to more data should, on average, make increased cognitive ability more valuable. Culturally acquired representations also allow people to learn things that they otherwise could not learn. For example, having words for numbers seems to make it possible to keep track of integers greater than three or four,[2] an ability that opens many new cognitive possibilities.

This reasoning suggests that there will be a positive feedback relationship between cognitive abilities and cultural learning. Unfortunately, the fossil and archaeological records provide little evidence about the nature of cognitive abilities and cultural learning in our hominin ancestors. About two million years ago, australopiths were replaced by species in the genus *Homo*. These creatures had brains that were about 50 percent bigger than those of the australopiths,[3] they made more sophisticated stone tools, and they substantially expanded their range across most of Eurasia. However, these creatures still developed almost as rapidly as apes, suggesting that cultural learning did not play a major role in their lives. Over the next 1.5 million years, brains got larger, but there is nothing that looks like cumulative cultural evolution in the stone tool record. However, the archaeological record is incomplete because it doesn't preserve information about perishable technologies—things like shelters, watercraft, plant detoxification methods, clothing, and cooking. Fast-forward to about three hundred thousand years ago, when the stone tool record becomes more modern, we see scattered hints of other technologies, and life history slows to close to the modern human pace. If forced to bet, I would wager that the advent of the genus *Homo* coincided with the evolution of increased cultural learning abilities, and that over the next million years, cultural learning abilities and cognitive abilities gradually ratcheted each other upward until a threshold was reached that allowed cultural accumulation. Given the pervasive uncertainty about both the archaeological record and the cognitive machinery underlying cultural learning, I am glad that I don't have to make this bet for real money.

Now a few words about our debts to previous scholars. Our project began when we were assigned to teach an introduction to environmental studies at the University of California–Davis. Our plan was to build the course around the theme that goal seeking by individuals leads to environmental outcomes that are bad for everybody. We knew that there were similar ideas about this in biology and economics. Pete was a devotee of George Perkins Marsh, so we wanted to cover pollution and overexploitation of resources in preindustrial societies. But when we looked at the then-fashionable literature on ecological anthropology, we weren't satisfied with what we found. Being young and cocky, we thought we could do better. The idea of cultural evolution was somehow in the air, and we read papers by Ted Cloak and Eugene Ruyle, and an excellent book by Ron Pulliam and Chris Dunford. We were strongly influenced by a paper on cultural evolution by the psychologist Donald T. Campbell, and we did some initial work using methods from evolutionary ecology, especially those used in Richard Levins's work. The papers of Luca Cavalli-Sforza and Marc Feldman opened our eyes to the power of models from population genetics and epidemiology. Luca and Marc generously allowed us to sit in on a course at Stanford based on their 1981 book; we also learned as much about population genetics as we could in courses taught by Michael Turelli and John Gillespie at UC Davis, and I was lucky to spend a year in Mike Wade's lab at the University of Chicago during the quantitative genetics renaissance. Our research style was, and still is, to read widely in anthropology, psychology, and economics looking for promising empirical problems and then tackle those problems with theory derived mainly from population biology.

We are aware that there have been important precursors in the social sciences. As Orr correctly notes, there are parallels in the work of the Austrian-school economist Friedrich Hayek. But after prospecting in Hayek's work a bit, I decided that the basic structure of his thinking was sufficiently different from ours that my time was better invested in pursuing our own research program. Much the same goes for the late-nineteenth-century sociologist Gabriel Tarde. It is quite possible that we have reinvented results and insights already gained by Hayek, Tarde, and others. We leave it to future scholars to assign credit where credit is due.

KIM STERELNY

One of my life rules is to never argue with philosophers. I learned this while trying to argue about levels of selection with Bob Brandon, a philosopher of biology at Duke. This was back in the early 1980s, and in those days I knew as much about the theory as anybody. But I still lost every argument. Subsequent experience has reinforced the lesson: don't tangle with professionals. However, I disagree with much of what Kim Sterelny says about my essay and want to explain why I think he's wrong.

First, there is one thing that Sterelny and I do agree on. There should be no doubt that collective intelligence always depends on individual intelligence. Even simple tasks like walking down a crowded city street require computations that are (for the moment) beyond the most sophisticated machine learning algorithms. Of course, this is just as true of butterflies and baboons as it is of human beings. What about what Sterelny refers to as "our distinctive human intelligence"? It is not clear what he means by this. Whether or not there is some useful measure of general cognitive ability that ranks us above other species is a vexed question, but I think there is evidence that people are better than other primates at specific tasks like causal reasoning and theory of mind that are plausibly related to our tool-making abilities.[4] It seems likely that these dimensions of human intelligence are an essential ingredient in our success. Sterelny seems to think that I believe that culture alone is sufficient to explain human success, and that if chimps were endowed with accurate social learning, they'd soon be making atlatls and Levallois points. But this is not what I believe. Many aspects of the human phenotype are crucial—cognitive abilities, bipedal locomotion, fine motor control of the hands, a life history with a long childhood followed by a rapid growth spurt—to name but a few. All these things are essential, but they are not enough. Cultural adaptation is also an essential ingredient. Sterelny is clear that he agrees with this claim, so no argument there.

Sterelny claims that I believe that cultural transmission is "largely the operation of a simple conformist heuristic." I assume that by "conformist," Sterelny means that the probability a learner acquires a trait is an increasing function of the number of people the learner observes using the trait,[5] and that other factors like individual experience and evolved psychological predispositions[6] do not play a significant role. If this is what he means, then he is mistaken—I do not believe any such thing.

First, as Sterelny notes, I think that learning biases that are focused on the properties of models like "copy the successful" are important in cultural learning. This means that cultural learning cannot be described by a simple conformist heuristic. A tendency to copy the successful will cause learners to disproportionately learn traits that lead to success, and as a result adaptive traits will spread even though learners have no causal understanding of why those traits generate success. There is much evidence from studies of dialect evolution, diffusion of innovations, and experiments with laboratory cultures that cultural learners use a success-based heuristic.[7] There is also evidence for a number of other learning mechanisms that increase the chance that learners will attend to some people more than others—people similar to the learner, people who exhibit credibility-enhancing displays, and so on. All of these mechanisms structure cultural learning so that learners are more likely to acquire adaptive traits. You can't have it both ways; if you accept that "model-based biases" are important, then you cannot claim that cultural transmission is described by a simple conformist heuristic.

Second, many aspects of our cognitive machinery shape the way we learn and so bias cultural learning, making it more likely that we learn some things and less likely that we learn others. I did not spend much time discussing this issue in my essay, but Richerson and I have dealt with this at length elsewhere.[8] Some of this cognitive machinery probably exists because it biases cultural learning in order to increase the probability of acquiring adaptive behaviors. For example, Clark Barrett has shown that children are more likely to retain knowledge about which animals are dangerous than information about which ones are edible,[9] probably because it is more important for them to learn about danger, and Annie Wertz has shown that they are also biased against eating vegetables, probably because, like nardoo, many plants are toxic.[10] The absence of such "content"[11] biases in the mathematical model described in my essay is an artifact resulting from the assumption that both environments were equally likely and the fitness effects were symmetric. Relax either of these simplifying but unrealistic assumptions, and the model predicts that learning will be biased so that learners are more likely to acquire the trait that is more likely to be adaptive on average.[12] Note that such biases channel cultural evolution into adaptive directions but by themselves are not usually sufficient to generate the complex cultural adaptations that have made people so successful. Being predisposed to learn that plants are toxic does not tell you to avoid nardoo—that requires local knowledge. It's also the case that much cultural learning is biased by cognitive machinery that exists for other reasons. Writing systems are only a few thousand years old, and until a few hundred years ago most people were illiterate, so the brain systems that allow us to read were not structured by natural selection because they allowed us to read. Nonetheless, these neural systems place many constraints on the nature of writing systems.[13]

Sterelny is also mistaken when he says that I think that "agents are rarely well placed to directly sample the environment reliably at low cost." Remember that in the model discussed in the essay, the accuracy of the environmental signal is a parameter that can be varied. My coauthors and I show that when the signal is accurate, selection favors learning rules that mostly ignore social information. So for behaviors for which the environment provides ample feedback, the model predicts that social information will be secondary, just as Sterelny says. The model predicts that people will mainly copy others only when it is hard to learn the best behavior. There is no doubt that life is full of situations in which individuals get lots of feedback from the environment and can adjust their behavior accordingly, and many of these are important for human adaptation. Louis Liebenberg's wonderful books on tracking provide excellent examples. It's also true that many skills must be refined through prolonged practice. Fijians rely on the *isele levu*, a long machete-like knife, for many tasks including house building. Learning how to use an isele levu to efficiently cut posts and beams and trim thatch requires lots of practice. As I can testify, being shown what to do is not enough. Sterelny is right that there are many examples of easy-to-learn traits and traits that require practice, but this is beside the point.

My claim is that there are many traits that are crucial for success in every human environment for which environmental feedback is rare or noisy, and natural selection has shaped human psychology so that learners attend to what other people are doing when learning such traits. In acquiring these traits, people use the commonness of a trait as a cue—Sterelny's conformist heuristic—but also exercise content and model biases. You can't disprove this claim by enumerating examples in which people get lots of environmental feedback. You have to show that traits with little environmental feedback are rare or otherwise unimportant for human adaptation, or that people don't use social information when they acquire such traits.

Sterelny suggests that if people commonly relied on social information, there should be lots of examples of maladaptive cultural traits, and he doubts that this is the case. This would be correct if there were no model biases, but that is not my claim. Rules like "imitate the successful" will tend to cull maladaptive beliefs, and the more maladaptive the belief, the more likely it is to be eliminated. This is tricky because, like natural selection, the strength of a model-based bias depends on the amount of variation in the population. When most people do the same thing, model-based biases are weak. This means that if a maladaptive practice becomes very common for some reason, it may persist for some time. So the model predicts that we should observe maladaptive traits, but they should not be too costly unless they are widely shared.

There are many, many examples of maladaptive culturally transmitted traits. Even Sterelny acknowledges that people have acquired lots of untrue religious beliefs but conjectures that these may not be very costly. In contrast, I think that religious devotion leads to celibacy, self-denial, and other forms of costly behavior. As Sterelny notes, systems of folk medicine seem to contain a lot of useless or harmful nostrums. In many places disease and misfortune are attributed to the malicious actions of friends and neighbors, and such beliefs often lead to very costly actions. For example, among the Gebusi, a group living in the Fly River region of Papua New Guinea, all deaths are thought to be the result of sorcery. When someone dies, a homicide inquest is held, and in a substantial fraction of cases someone is found guilty and executed.[14] Despite great wealth, most people in the industrial world have very small families and do not maximize their lifetime fitness.[15] The list of plausibly maladaptive behaviors is long.[16]

There are also probably many practices that are useful, but not optimal. They persist because they are socially acquired and then not rigorously tested. People rely on complex artifacts that present difficult design problems. The Fijian houses discussed in the text have many different attributes, and builders have to make many decisions about materials, structural details, and dimensions. As Sterelny notes, there is feedback about some attributes like the thickness of the thatch. Does the roof leak when it rains? But even this is not completely obvious, because the cost and availability of thatch vary over time, and a roof that leaks a little may be optimal if thatch is costly. Structural design is important during hurricanes, which may not occur for years, and when they do, luck also plays an important role in determining which houses survive. And these houses are not unusually complex. People in many small-scale societies depend on equally complex tools—kayaks, composite bows, boomerangs, and on and on. And many technologies depend on causally opaque processes. Manioc (also called cassava) is a highly productive tuber widely used in tropical South America. But some varieties contain significant levels of cyanide, and South Americans use a number of different techniques to detoxify this plant.[17] These techniques typically involve a number of steps spread over several days and are very labor intensive. It seems unlikely that people understand the contribution of each stage to the overall level of detoxification.

So how can we know that such complexities lead people to adopt suboptimal practices? One source of evidence for this is that culturally distinctive people in similar habitats make different artifacts. Traditional Tongan houses have a different design than Fijian houses. Of course, it's possible that there is some subtle environmental difference that makes Tongan houses better in Tonga and Fijian houses better in Fiji, but this doesn't seem likely. The basic Fijian design is used widely on islands within the Fijian archipelago that vary in hurricane frequency, forest cover, and other factors that should affect house design. This suggests that the design does not depend on details of the habitat. The eastern islands of the Fiji archipelago were part of the Tongan maritime empire for much of the last millennium, and on these islands Fijians build Tongan-style houses. This suggests that house design is usually acquired along with other parts of the Fijian cultural package. Similar evidence exists for other cultural traits. Manioc spread from the Americas to Africa after European contact, but techniques used for processing manioc in the Americas did not, suggesting that these techniques are not easily learned using environmental cues.[18] More generally, the fact that a people's language predicts their technology and subsistence practices as well as or better than characteristics of their environment indicates that technology and subsistence practices are not being optimized on generational timescales.

Manioc processing illustrates why the reliance on cultural cues can be a powerful mechanism expanding the ability of human populations to adapt. It appears that the processing techniques are hard for individuals to learn because they are complex and causally opaque, and the environmental feedback is poor because it sometimes takes a while before the costs of eating these foods unprocessed becomes apparent. If people adopt only those practices whose benefits can be understood, they will not end up with an effective detoxification procedure. On the other hand, if they rely on social cues, populations can gradually evolve practices that greatly expand their subsistence base.

Sterelny has also misunderstood the relevance of the manufacture of complex artifacts by beavers, termites, amoebas, and the like. Of course they are not examples of learning. That is exactly the point. Cognitive psychologists like Alison Gopnik have argued that causal understanding is necessary because without it artifacts cannot be modified in response to local environmental contingencies. The artifacts produced by other animals prove that accurate genetic transmission plus a dumb process like natural selection can generate algorithms that allow the construction of highly designed artifacts that vary in response to contingencies without any insight on the part of the builders. This supports the argument that accurate cultural transmission plus dumb processes like "imitate the successful" could do the same.

RUTH MACE

Ruth Mace and I have tangled a bit in the past, but I am pleased that I agree with most of what she says here. I was especially gratified by her observations on the state of evolutionary social science. Disciplinary ethnocentrism is the bane of the human sciences. Undergraduates take classes in economics, anthropology, and psychology and are told different and incompatible things about human behavior and society. Given the difficulty of understanding ourselves, this is perhaps forgivable. But it is unforgivable that many scholars are completely comfortable with this state of affairs. For many years, evolutionary social scientists re-created a mini-version of this balkanization in which members of the three main camps—human behavioral ecologists, evolutionary psychologists, and cultural evolution researchers—hardly talked to each other. To our credit, however, I don't think that very many of us were comfortable with this situation. A shared evolutionary perspective demanded that we resolve our differences. And like Mace, I think that this has begun to happen over the last several years. Mace's own work using phylogenetic methods to study variation in human behavior is a prime example of such a synthesis, and many researchers, especially the younger ones, move easily from psychology to population dynamics to adaptation.

In what follows I will discuss two ideas about the linkage between cultural evolution and human behavioral ecology that arise from Mace's discussion of my paper.

The first idea is that cumulative cultural adaptation changes the scale on which adaptation takes place, and that human behavioral ecology, as it is usually practiced, depends on the fact that cultural adaptation allows different human populations to evolve different, local adaptations. As Mace notes, human behavioral ecologists try to account for behavioral variation among human populations. For the most part, predictions are based on models that assume that people maximize genetic fitness in their local environment. Human behavioral ecologists usually apply fitness-maximizing ideas without worrying too much about whether the variation is due to evolved genetic differences, cultural adaptation, or individual learning. This gambit is rooted in the idea that natural selection will tend to shape developmental programs so that outcomes are fitness maximizing whether or not they are sensitive to environmental contingencies. Behavioral ecologists are well aware that proximate mechanisms sometimes matter but think that fitness maximization is a powerful heuristic. And I think that experience in biology proves them right. I also agree with Mace that cumulative cultural adaptation often produces fitness-maximizing human adaptations.

Human behavioral ecologists sometimes characterize their work as just ordinary behavioral ecology applied to humans. But this is not really correct. Behavioral ecologists who study other species focus mainly on differences among species, not variation among populations of the same species. This is not to say there is no variation in behavior across populations. For example, we know that chimpanzees in some parts of West Africa use stones to crack open hard-shelled nuts, while chimpanzees in other populations do not. But the magnitude of this variation is modest compared to the variation among species. In contrast, there are huge behavioral differences between human populations that cry out for explanation.

I believe that extensive differences between human populations are possible only because humans adapt culturally. Without cultural adaptation, our species would be like all the rest. Learning, reasoning, and other forms of individual adaptation would allow modest differences to develop between populations, but cognitive limitations would prevent the extensive local adaptation that characterizes different human populations. Some human behavioral ecologists (not Mace) are skeptical about the importance of culture in creating variation. As a noted human behavioral ecologist once said to me after reading *Culture and the Evolutionary Process*, "too much attention to a little bit of development." But if I am right, the intraspecific variation that human behavioral ecologists study depends on cultural adaptation.

Mace's own research beautifully illustrates this claim. Behavioral ecologists want to identify the ecological causes of differences between human groups. For example, does a pastoral economy cause people to adopt patrilineal inheritance? To answer such questions, Mace collects data on many human groups and determines whether pastoralism and patrilineal inheritance are statistically associated. Anthropologists have long known that in doing this you have to somehow correct for the fact that groups that have a shared cultural history are not independent data points, but they lacked systematic methods for making this kind of correction. Similar problems exist in comparative biology, and biologists have developed sophisticated statistical methods to deal with them. Mace has applied these methods with great success to explaining cultural variation. This work implicitly relies on the premise that human populations have historical trajectories that are analogous to the phylogenetic histories of nonhuman species. If people adapted only by learning or individual calculation, there would not be any phylogenetic inertia to correct for, and phylogenetic methods would be unnecessary.

Now I want to turn to Mace's challenge about how to test hypotheses based on cultural group selection. I definitely agree with Mace that the theory of cultural group selection is well worked out, and I think there are a number of plausible empirical examples of cultural group selection, but I also agree that more systematic empirical testing is needed. However, the tests have to be done right. And here, I want to disagree with one of Mace's suggestions.

Mace proposes that individual benefits, not cultural group selection, explain why Turkana men participate in warfare. But this is not an alternative hypothesis at all. Cultural group selection models assume that behavior within groups is motivated by individual self-interest. Norms are maintained by rewards and punishments that make it beneficial to follow the norms. If individuals did not benefit from conforming to norms, then the cultural group selection hypothesis would be falsified. Think about the Turkana man who goes along on a raid and keeps his head down when the shooting starts. When the time comes to divide the loot, it's not in any individual's private interest to prevent him from taking a share in the absence of a mutually enforced norm. But when such norms exist, it is individually advantageous, on average, to conform to the norms. Individual advantage explains why people obey norms and why people enforce norms, but it doesn't explain the content of norms. Why, for example, do the Turkana condemn raids on other Turkana but valorize raids on members of other tribes? The cultural group selection hypothesis proposes that norm content—why people have the norms they have—is explained by competition between groups in which individual advantage stabilizes different norms.

To put it another way, cultural group selection models do not predict that parochial altruism will evolve.[19] Mace describes empirical work in Northern Ireland that indicates that intergroup conflict does not engender altruistic behavior in experiments. This is very interesting work because other researchers have observed the opposite result, and Mace's result counts against hypotheses that suggest we have a genetically evolved predisposition to cooperate more when threatened by outsiders. But Mace's data have no bearing on cultural group selection one way or the other because cultural group selection predicts that norms maintained by self-interest will tend to be group beneficial, not that people will be altruists. Moreover, cultural group selection models predict that the group boundaries that matter will be those that separate culturally different groups. It seems very unlikely that these boundaries would define Belfast neighborhoods and would instead predict that norms should benefit much larger entities like all Catholics or all Protestants.

I agree with Mace that it is important to go beyond inference from design when testing cultural group selection models. Some empirical work of this kind does exist,[20] but it is not definitive. But I do not agree that this by itself should cause us to reject cultural group selection hypotheses. Science should always be seen as a competition between contending hypotheses. You can't test the hypothesis that cultural group selection has shaped norm content unless you have some competing hypotheses. And here human behavioral ecology comes up short. As far as I can see, human behavioral ecologists have no plausible hypothesis explaining norm content. Mace suggests that coercion by leaders explains why people have the norms they do, but this can't be right. Even in the simplest foraging societies, norms are shared by sizable numbers of people, and in food-producing groups like the Turkana, tens or hundreds of thousands of people share norms. In such large groups, it would be very costly for individuals or even small coalitions to generate sufficient coercive power to motivate individuals to adopt and enforce new norms. Moreover, in the nonstate societies that have dominated human political organization until very recently, the scale of cultural variation is usually much larger than any political institution that could make decisions about norm content. Economists have models of the evolution of norm content that are based on individual optimization, but these do not produce group-beneficial outcomes and are implausibly slow. Absent some plausible competing hypothesis, I think we should keep the cultural group selection models in play.

PAUL SEABRIGHT

Paul Seabright argues that my account of norms neglects three features that make norms less effective than I claim. These are that people often disagree about whether a norm has been violated, they disagree about what norms should be enforced, and conflicts of interest lead people to manipulate the process of norm enforcement to their own advantage. Seabright is surely correct. These are all real impediments to the maintenance of cooperation. It's hard to tell whether he thinks they are deal breakers. On the one hand, he paints a very dark picture that might lead you to think that informally enforced norms can never work. On the other, he accepts that informally enforced norms do explain cooperation among the Turkana.

So let me be clear about what I think. Seabright's problems do not create insuperable difficulties as long as groups are small and culturally homogeneous. Informally enforced norms cannot maintain cooperation in large, culturally heterogeneous societies, and the large-scale cooperation that we see in complex societies requires the institutions (which are really novel and more complex sets of norms) that have evolved over the last ten thousand years or so. But informally enforced norms make cooperation on the scale of forager tribes and small horticultural villages possible. In what follows, I will explain why monitoring, norm communication, and conflicts of interest are not such big problems in small-scale, culturally homogeneous societies. I will then discuss some crucial differences between the ways Seabright and I think about cultural transmission, and why Seabright's picture leads him to overestimate the importance of the dark side.

Let's start with monitoring. Seabright is right that it has been a long time since I have had to figure out which of my kids was the culprit (although my youngest is about the same age as his, so it's not clear that he is an expert either). However, this misses the point. Adults are trying to impose *their* norms, and they don't live in the children's world and don't know what's going on in that world. Like adults, children live in a world regulated by norms. When my son was six, we returned to Los Angeles from a remote field site where he had interacted with only two other children for almost an entire year. A few weeks after our return, he came back from first grade looking a bit down. When I asked him what was wrong, he sighed, "There are so many rules." Further conversation revealed that he didn't mean rules imposed by the teachers—those were easy. It was the myriad unstated rules that governed the schoolyard that gave him trouble. Children monitor each other and enforce norms among themselves, sometimes ruthlessly.

The scale of life in the village is more like life in the schoolyard than life in modern, anonymous, urban, multiethnic societies. The world is small, much of everyday life is out in the open, people have lots of opportunity to find out what's going on, and they are intensely interested. When you encounter someone as you walk through the Fijian village in which I have worked, a standard greeting is "Where are you going?" As an urbanite, my first reaction to this query was to think "None of your business." But for the villagers, that reaction would be rude—what you are doing *is* everybody's business. In small-scale societies, people know a lot about what is going on, and they share this knowledge. It's not perfect. People get away with stuff, there are secrets, people carry on affairs, and thefts occur. There are also disputes about who did what, and normative mechanisms for resolving such disputes. Despite all this, people know that there is a good chance they will be caught if they do something wrong. The wisdom of crowds works best when crowds are small and intimately acquainted.

To see how this works, let's think about norms that enhance cooperation in the everyday lives of foraging peoples. Food sharing, particularly of meat, plays a crucial role in hunter-gatherer economies because it acts as a kind of insurance policy, reducing the risk associated with hunting large game. In many groups, norms specify who gets which cut of meat from kills. For example, we saw earlier that among the Mbendjele of the Congo Basin, the hunter gets the heart, other hunters get the kidneys, and so on. Among some Kalahari groups, the hunter gets only the ribs and one shoulder blade, and young men can eat only the kidneys, abdominal walls, and genitals. In many groups, the hunter is not allowed to make the distribution. Among the famous Ju/'hoansi of the Kalahari, it is the man who made the arrow that killed the animal, not the hunter of the animal, who divides the kill.[21] If you think about an old man butchering an impala in a desert camp, surrounded by people eager for their just share of the meat, I don't think you will find it so implausible that people who take more than their share will be detected. You might think that people would cheat by consuming their kill out in the bush so that nobody would know. This does happen, but according to my ASU colleague Kim Hill, among Aché foragers, such norm violations are often detected because the Aché are superb trackers who, in a kind of forest CSI, figure out who did what from evidence left at the scene. It's also interesting that in both the Congo and the Kalahari, norm compliance is buttressed by the shared cultural belief that norm violators will experience bad hunting luck in the future, a hard-to-falsify incentive to behave properly.

Seabright suggests that it is easier to explain cooperation in public good settings because people have a shared interest. I think public good provision is the hardest case to explain because groups are large and reciprocity can't work. Be that as it may, notice that food sharing is not a public good, and there are sharp conflicts of interest among participants. Young men, who could use their strength to monopolize the kill, are disadvantaged, as are better-than-average hunters. Sharing provides a long-term benefit, and you might think that norms would not be needed to sustain such small-scale cooperation. However, while both related-ness and reciprocity do play a role in hunter-gatherer food sharing, the evidence is that they are not enough—norms also play an important role.[22] Also remember that other pri-mates share very little, and when chimpanzees do make a kill, other animals crowd around the hunter pushing and begging.

I agree with Seabright that consensus about norms is also important. When people don't agree about what is right and wrong, there is going to be trouble. As Seabright points out, the conflicts we see in the modern Middle East are a depress-ing example of this. But this doesn't mean that intentional communication of normative rules is important. In fact, the intentional communication of norms plays a modest role in the transmission of norms in face-to-face societies. For Seabright, "Cultural transmission is a strategic exercise in which those who transmit information seek (not always suc-cessfully) to manipulate those who receive it." This picture of cultural transmission is common in economics,[23] but it is in-consistent with the evolutionary approach to social learning outlined in my essay, and more importantly, with the empiri-cal evidence about how cultural transmission actually works.

First and foremost, cultural learning is structured mainly by receivers, not transmitters. Children are not empty vessels into which adults pour cultural information. They are born with a mind adapted to extracting useful knowledge from their social environment. In most societies, young children first learn from their parents and other family members be-cause these are the people who are readily available. Then as they get older, kids spend lots of time outside the household and can observe what is going on in the village, and they learn what people around them are doing. Language learning is an example. Kids learn thousands of word meanings by eavesdropping on conversations. Teaching is probably important, but not because somebody is trying to convince them to use "dog" instead of *hund*. Instead, adults and kids engage in joint attention, and this helps kids narrow down the range of hypotheses that they have to entertain about what words mean. The same goes for lots of adaptive skills. People are doing what they are doing, and children are learning how to do those things by watching, and then participating.[24] A young girl sees that all of the mat makers use pandanus leaves that are harvested at a particular time and cured in a particular way, and they weave the mats using a particular technique. Teaching helps learners mainly by pointing out what is important and also helps them acquire difficult manual skills like weaving. Children are immunized against being manipulated because they can observe what people are doing in their everyday lives and check what they say against what they actually do. If children see that women don't eat big predatory fish when they are pregnant, despite the costs of forgoing the nutritious meal, then they are likely to believe that women think that these fish really are harmful. There is a risk of manipulation with beliefs that are learned through verbal instruction. However, this risk can be minimized by using the "put your money where your mouth is" principle. Someone tells you that a failure to follow sharing norms will lead to bad luck in future hunts. If the teacher is some old man who never hunts, you should be suspicious, but if it is a successful hunter who scrupulously adheres to sharing norms, the risk is smaller. There is both laboratory and field evidence that learners are more likely to adopt the behavior of people who "put their money where their mouth is."[25]

This means that people in culturally homogeneous societies do not very often choose among alternative norms. They adopt the norms that govern their social world. Fijians believe it is rude to be higher than another person, and since much social interaction takes place when people are sitting cross-legged on the floor, entering or moving in a house is done on hands and knees. Children learning norms see this thousands of times during their childhood. They aren't told the norm; they observe it in action. The same goes for more important norms. Fijians have a strong norm against any kind of social interaction, including speaking, between certain kinds[26] of cousins of different genders. Once again, children see many examples of this norm in action, as well as occasional inadvertent violations and the resulting social opprobrium. For the most part, norms aren't communicated or transmitted; they are adopted.

The spatial and temporal scales on which norms vary support this picture. Versions of the Fijian norms I just described are found in villages throughout the Fijian archipelago. There were roughly three hundred thousand people in Fiji when they were contacted by Europeans. The decisions of an individual cannot explain the content of such widespread norms. Seabright suggests that norm content is arrived at through some kind of political process. But the scale on which norms are shared in small-scale societies is typically much larger than the scale of any political institution within these societies. Among hunter-gatherers, there are typically no permanent institutions larger than the band, and even in reasonably complex societies like the ones of the Fijian islands before colonization, polities typically numbered a few thousand people. Polynesian societies that have been separated for at least one thousand years and are spread across the eastern Pacific also share related norms, including a political system based on ranked patrilineal clans.

Exceptions to this picture occur when, for some reason, there is rapid technological or environmental change leading to rapid change in norms. The introduction of the horse to the Great Plains in the sixteenth and seventeenth centuries led to striking changes in the societies of Great Plains Indian groups. Foragers like the Comanche, corn farmers like the

Crow, and boreal forest foragers like the Cheyenne moved to the plains and converged on a new social system based on large summer encampments regulated by "men's societies," frequent warfare, and painful summer initiations.[27]

I agree with Seabright that conflict among norms is an old problem. It's not right to think of norms as simple rules like "thou shalt not kill." Much ethnographic evidence suggests that there are always contingencies.[28] Were you defending yourself? Were you drunk? Were close kin victimized? Was the victim doing something that impugned the family honor? Real norms are complicated, nuanced algorithms that have rules for dealing with such contingencies.[29] But even then, norms sometimes come into conflict. As long as such conflicts are not too common, the system can work. Unlike Seabright, I am not keen on folktales as evidence about past societies. The problem is that the content of such tales is not determined by some social function, but instead by what people find interesting and memorable,[30] and there are good reasons to suspect that this makes the content of folktales biased in favor of situations that involve conflicts. As an analogy, imagine that sometime in the future, somebody tried to reconstruct twentieth-century English life from a collection of Agatha Christie novels. They'd get some things right but would vastly overestimate the likelihood of murder.

Notes

CHAPTER 1: NOT BY BRAINS ALONE

1 Technically, sporocarps, seedlike structures that produce spores.
2 Howitt et al. 1862; Cathcart 2014; Moorehead 1963
3 Nowak 1999, 664
4 Aplin et al. 2003
5 Species biomass is the sum of the masses of all individuals belonging to that species at a given time.
6 Hill et al. 2009
7 Binford 2001
8 Kay and Hoekstra 2008
9 Collias and Collias 1964
10 E.g., Klein 2009
11 Wadley et al. 2009
12 Ibid., 9593
13 Tennie et al. 2009; Dean et al. 2014
14 Collias and Collias 1973
15 Tebbich et al. 2001
16 There are many examples of the lost European explorer experiment. The Franklin expedition is discussed in Boyd et al. 2011a, and a number of other examples are given in Henrich 2015.
17 Howitt et al. 1862; Cathcart 2014; Moorehead 1963
18 Cathcart 2014, Kindle location 2158–60
19 Earl and McCleary 1994
20 Clarke 2012, Kindle location 1029
21 Ibid., Kindle location 6584
22 Wadley et al. 2009, 9593
23 Tooby and DeVore 1987; see Barrett et al. 2007 and Pinker 2010 for updates of this argument.
24 The view that people adopt behaviors because they understand why they are better than alternatives is widespread in the social sciences. In economic models, actors innovate at a cost, and other actors adopt the innovations because they understand how they work and why they are beneficial. See Arrow 1962 and Romer 1993 for canonical examples. Many psychologists and philosophers take a similar view, and it is a common folk model as well.

25 Dawkins 1986
26 Henrich and Henrich 2010
27 Broesch 2010
28 Boyd and Richerson 1987, 1995, 1996; Perreault et al. 2012
29 See Perreault et al. 2012 for details.
30 We assume that the population is large so that we can ignore random changes in the composition of the population due to sampling variation.
31 There are two averages to keep track of. The first is the average value across individuals in the population at any time. This average will never stop changing, because the environment switches on average every T generations. After a switch, many individuals will behave maladaptively, and large values of g will be favored. As the fraction of individuals who behave correctly increases, smaller values of g are favored. The second average is the average over a long enough time period that many environmental switches have occurred. This average value of g will settle down to a constant value, which is the value favored by selection over the long run.
32 Denham et al. 2009
33 E.g., Boyd and Richerson 1985, 1996
34 Basalla 1988; Petroski 1992a
35 S. Johnson 2010
36 See Henrich and Gil-White 2001 and Henrich 2015 for extensive reviews of this evidence.
37 See Richerson and Boyd 2005 and Henrich 2015 for more examples and data from experiments.
38 Henrich 2009
39 See Henrich 2015 for a review.
40 Hansell 2005
41 Collias and Collias 1964
42 Gould and Gould 2007; Hansell 2005
43 Jacklyn 1992
44 Gould and Gould 2012
45 Enquist et al. 2008
46 Perreault 2012
47 Gingrich 1983
48 Lane 2009. In the case of striated muscle, it may be twice; see Steinmetz et al. 2012.
49 Lane 2009, Kindle location 1174
50 Tennie et al. 2009
51 Boyd and Richerson 1996
52 Whiten et al. 2005
53 Hill et al. 2014
54 Gergely and Csibra 2006

55   T. Morgan et al. 2015
56   Boyd and Mathew 2015
57   Boyd and Richerson 1996
58   Mathew and Perreault 2015
59   Mathew and Perreault used principal components methods to reduce the dimensionality of the predictor variable so that the number of predictor variables in each category was comparable and summarized the same amount of variation in predictors. They used a model comparison statistic to choose the best logistic regression model for each trait and then summed the standardized beta values for cultural history, ecology, and distance to generate an aggregate measure for each.
60   Jordan 2015
61   Currie et al. 2010
62   Holden and Mace 2003
63   Shennan 2001
64   Henrich 2004, 2006
65   Powell et al. 2009
66   Kline and Boyd 2008
67   Derex and Boyd 2015
68   Cosmides and Tooby 2006
69   See Richerson and Boyd 2005, or Henrich 2015, for more details.
70   Beppu and Griffiths 2009
71   Boyer 2001
72   Henrich and Gil-White 2001; Henrich 2015
73   Henrich and Gil-White 2001; also see Boyd and Richerson 1985, ch. 8.

CHAPTER 2: BEYOND KITH AND KIN

1   You can watch this on YouTube. See https://www.youtube.com/watch?v=R5Gppi-O3a8.
2   Petroski 1992b
3   Seabright 2010
4   Greenhouse 2014
5   Kaplan et al. 2000
6   Hill 2002
7   Ibid.
8   Wood and Gilby, forthcoming
9   Lukas and Clutton-Brock 2012a, 2012b
10   Watch it on YouTube: https://www.youtube.com/watch?v=0AMoBU7lFUA.
11   The Pleistocene epoch began 2.6 million years ago and ended 11,700 years ago.

12 B. Smith 2011
13 Bliege-Bird et al. 2009
14 B. Smith 2011
15 Clastres 1972
16 Steward 1933
17 Wilke 2013
18 Brink 2008
19 Brink 2005; Friesen 2013
20 Brink 2013; O'Shea et al. 2013
21 Swezey and Heizer 1977
22 Gat 2015
23 E.g., Fry 2006; Manson and Wrangham 1991
24 Gat 2015; LeBlanc 2014
25 J. Morgan 1852
26 Gat 2015
27 Sutton 2014
28 M. Smith 1938; Hämäläinen 2008
29 Bamforth 1994
30 Darwent and Darwent 2014
31 James and Graziani 1992
32 Kroeber 1976
33 Darwent and Darwent 2014
34 Couzin and Franks 2003; Franks 1986
35 E.g., Sterelny 2012; Pinker 2010
36 Iwaniuk and Arnold 2004
37 This definition of r, b, and c is exact only when selection is weak and a number of other plausible conditions hold. See Birch and Okasha 2014 for a clear introduction.
38 To see why, suppose that helping to build a shared facility like a burrow or a drive line decreases a focal individual's fitness by one unit, but this individual's participation increases the fitness of each of the individuals in his or her local group by 1/10 of a unit. If the group is made up of 100 individuals, the net change in the focal individual's inclusive fitness is $0.1 \times 1 + 0.1 \times 99 \times r - 1$, where r is the average coefficient of relatedness between the focal and other group members. If group members are unrelated, $r = 0$, the change in inclusive fitness is $0.1 \times 1 + 0.1 \times 99 \times 0 - 1 = -0.9$, and selection does not favor participation. If the members of the group are full siblings, as is possible with insects but unlikely with mammals, then $r = 0.5$ and the focal individual's inclusive fitness is $0.1 \times 99 \times 0.5 + 1 \times 0.1 - 1 \approx 4$ and the participation is favored by selection.
39 Boomsma et al. 2011
40 Hill et al. 2011
41 Hill et al. 2014

42 Ibid.
43 Relatedness depends on group size and migration rate. Large group size and high migration rates reduce relatedness. To see why, think about randomly picking two members of a group. Now, trace their genealogies backward in time. One of two things can happen. You might be able to trace the genealogies back to a common ancestor, and if that is the case then the two individuals will be identical by common descent. However, you might find that one of the two lineages will lead to an immigrant who came from outside the group. Because the population is large and immigrants come from far away, we can be certain that an immigrant does not share a common ancestor with a group member. Relatedness is the probability that the two individuals have a shared ancestor. The larger the population, the farther back you will have to go to find a shared ancestor. And as the period of time increases, more migration events will occur. This makes it more likely that one of the ancestors will be an immigrant, and thus unrelated. High migration rates will increase the likelihood of this occurring. Thus, large group size and high migration rates will both reduce relatedness within populations.
44 Langergraber et al. 2011
45 Bowles 2006
46 Trivers 1971
47 Axelrod and Hamilton 1981
48 McElreath and Boyd 2007
49 Maskin et al. 2015
50 Clutton-Brock 2009
51 Boyd et al. 2014
52 For example, suppose that I fight as long as half my group cooperates. As long as I am in a group in which at least half of the others are cooperators, then some cooperation can be sustained, and under certain conditions populations with a mixture of cooperators and defectors can be evolutionarily stable. But with plausible ratios of benefits and costs, groups have to be very small. Moreover, this process will generate some groups that cooperate and some groups that don't. More-tolerant strategies also increase more easily when rare, but this effect is important only in quite small groups. See Schonmann and Boyd 2016.
53 Hart and Pilling 1960
54 Marlowe 2009
55 Panchanathan and Boyd 2004
56 E.g., Raihani et al. 2012
57 E.g., Panchanathan and Boyd 2004
58 Boyd et al. 2014; Schonmann and Boyd 2016
59 E.g., Marlowe 2009; Guala 2012

60   Mathew and Boyd 2011, 2014
61   Mathew 2017
62   Leimar 1997
63   See Richerson and Boyd 2005, ch. 6, or Henrich 2015 for a more extensive treatment.
64   Reviewed in Richerson et al. 2016.
65   Holden and Mace 2003
66   Mathew and Perreault 2015
67   Sosis 2000
68   Henrich et al. 2010
69   Sharif et al. 2010
70   Kelly 1985
71   Soltis et al. 1995
72   Henrich and Gil-White 2001; McElreath et al. 2008
73   Boyd and Richerson 2002
74   Stark 1997
75   E.g., Alba and Nee 2003; Martin 2005
76   E.g., Cronk 2002
77   The level of density-dependent competition within groups is also important, but for simplicity, I will ignore this issue here. See Lehmann et al. 2006 for a review.
78   See David Queller's contribution to the Edge debate for a clear and witty explanation of the equivalence of the two approaches: http://edge.org/conversation/the-false-allure-of-group-selection.
79   E.g., West et al. 2011
80   The analogous process is probably not important in genetic evolution, because selection is much weaker and cannot maintain differences in gene frequencies in competing groups. See Boyd et al. 2011b.
81   Social insect colonies are not a counterexample here, as in many cases their effective group size is two, the male and female reproductives who founded the colony.
82   See Boyd et al. 2011b.
83   West et al. 2011
84   Boyd et al. 2011b
85   E.g., Boyd and Richerson 1985, 1990, 2002
86   E.g., Gavrilets 1996
87   Zefferman and Mathew 2015
88   Appiah 2010
89   Ibid.
90   E.g., Young 2001
91   E.g., Barton and Rouhani 1993
92   Kaplan et al. 2000
93   Clutton-Brock 2009; Schino and Aureli 2009
94   Sugiyama 2003

95 Mathew et al. 2013
96 Meggitt 1965
97 Mathew et al. 2013
98 For an interesting evolutionary account of why we internalize norms even though this constrains our choices, see Henrich 2015.
99 Axelrod and Hamilton 1981; Lehmann and Keller 2006
100 Nowak and Sigmund 1993
101 Fudenberg et al. 2012
102 Panchanathan and Boyd 2004
103 Boyd and Mathew 2015
104 Mathew and Boyd, forthcoming
105 E.g., Stevens and Stephens 2003
106 Dal Bó and Fréchette 2014
107 Gurven 2004
108 Lewis 2014
109 Boyd et al. 2011b

CHAPTER 4: ADAPTATION WITHOUT INSIGHT?

1 Rob does not say much about the origin of innovations, and so I too will set that issue aside.
2 The model's parameters are, in fact, the actual reliability of the signal from the world, and the actual population frequencies of the different behavior variants. But to make a choice, the agent must represent that reliability and those frequencies.
3 Though it will be noisy data, as technology is not the only factor affecting success rates.
4 It is true at the level of proximate mechanism. It remains possible that the costs and benefits of first-order free riding differ, and its control differs systematically, from the costs and benefits of higher orders.
5 Or perhaps "has shaped"; it is not unreasonable to model Boyd and Richerson as a superorganism.
6 I thank the lecture organizers for inviting me to participate in the lectures and the discussion those lectures generated, and I thank Rob and my fellow commentators for an enjoyable and stimulating experience.

CHAPTER 6: ADAPTABLE, COOPERATIVE, MANIPULATIVE, AND RIVALROUS

1 See in particular Boyd and Richerson 1985; Mathew and Boyd 2011; and Henrich 2016.
2 Group selection as it is known to biologists occurs when behavior that is individually maladaptive is selected because it contributes

to the fitness of groups that are in competition with other groups. What Boyd calls cultural group selection occurs when behavior that is individually adaptive contributes to the maintenance of norms that are selected by competition between groups that maintain different norms.

3   "Beyond the Fringe: From the 'Aftermyth of War' Sketch," https://www.youtube.com/watch?v=Y5YW4qKOAVM.

CHAPTER 7: CULTURE, BELIEFS, AND DECISIONS

1   Scott-Philips 2014
2   Carey 2011; Pica et al. 2004
3   Kimbel and Villmoare 2016
4   Vaesen 2012
5   There is a lot of confusion about this term. Pete Richerson and I (Boyd and Richerson 1985) defined conformist transmission as a social learning rule that causes the more frequent trait in the population to increase in frequency and argued that such a rule acts to increase the probability of acquiring adaptive traits in variable environments. Several authors have explored this logic, and at present the jury is out (Henrich and Boyd 1998; Wakano and Aoki 2007; Perreault et al. 2012) about whether it has this effect. In experiments dating back to the 1950s, psychologists have used the term "conformist" to mean a tendency to imitate the common type, a practice recently endorsed by Cladière and Whiten 2012. With this definition, conformist transmission can increase the rare type in the population.

   Sterelny is not clear, but since I do not mention conformism in the essay, I assume he means the latter definition. The ESS transmission rules shown in figure 1.5 of the essay are conformist only for high frequencies of the favored trait.

6   Cladière and Sperber 2007
7   See Boyd and Richerson 1985; Richerson and Boyd 2005; and Henrich 2015 for examples and references. Derex and Boyd (forthcoming) show that success bias leads to the spread of high payoff traits without the spread of causal understanding in an experiment using laboratory cultures.
8   For example, Boyd and Richerson 1985, ch. 4; Richerson and Boyd 2005, ch. 5
9   Barrett and Broesch 2012
10   Wertz and Wynn 2014
11   Henrich and McElreath 2003
12   See Boyd and Richerson 1985 for models of the evolution of such biases.

13   Dehaene 2009

14   Knauft 1985

15   Kaplan et al. 1995

16   See Richerson and Boyd 2005, ch. 6 and 7. Also see Henrich 2015 and Edgerton 1992 for more examples.

17   Henrich 2015

18   Ibid. Urban Venezuelans have also experienced manioc (there called yucca) poisoning due to ignorance of the necessary processing techniques. See Nicholas Casey, "No Food, No Medicine, No Respite: A Starving Boy's Death in Venezuela," *New York Times*, December 25, 2016, http://www.nytimes.com/2016/12/25/world /americas/venezuela-hunger.html.

19   Cultural group selection may have created social environments in which ordinary natural selection acting on genetic variation favored parochial altruism on the same scales as cultural variation. See Richerson and Boyd 2005, ch 7.

20   Richerson et al. 2016

21   See Henrich 2015, ch. 9 for an excellent discussion of norms in foraging societies, and for more details about Kalahari forager norms.

22   Jaeggi and Gurven 2013

23   See, for example, the influential work of Bisin and Verdier 2001.

24   See Kline et al. 2013 for supporting data.

25   Henrich 2009. It is important to see that credibility-enhancing displays are not stabilized for the same reason as costly signals as these are understood in biology and economics. The signaler is never paid back for the cost of the costly behavior. Instead the behavior is more likely to be acquired by others, something that need not benefit the signaler.

26   Specifically, what anthropologists call patrilateral parallel cousins— these are your father's brother's children.

27   Oliver 1962

28   Edgerton 1985

29   Mikhail 2007

30   Zipes 2012

NOTES TO FIGURES

Figure 1.1 Henrich and Henrich 2010

Figure 1.2 From Henrich and Henrich 2010

Figure 1.5 Redrawn from Perreault et al. 2012

Figure 1.6 From Collias and Collias 1964

Figure 1.7 From Hansell 2005

Figure 1.8 Photo by Neil Liddle.

Figure 1.9 From Perreault 2012

Figure 1.10 The relative importance of cultural history and ecology was calculated as follows. For each behavioral trait, multivariate logistic regression was used to estimate the importance of predictor variables in explaining variation in trait values. There is a beta coefficient for each pair of predictor variables and behavioral traits. Larger (in absolute value) beta coefficients indicate that a given predictor variable better predicts the occurrence of the behavioral trait. Predictor variables were categorized as cultural history, ecology, or spatial distance and then summed to give the overall importance of that category of predictor. An arcsine transformation was used so that the scales between zero and one and one and two are the same.

Figure 1.11 This graph plots the importance of cultural history (the sum of the absolute value of the beta coefficients of linguistic category for all traits). Redrawn from Mathew and Perreault 2015.

Figure 1.12 From Kline and Boyd 2008

Figure 1.13 From Derex and Boyd 2015

Figure 1.14 From Derex and Boyd 2015

Figure 1.15 Beppu and Griffiths 2009

Figure 1.16 Beppu and Griffiths 2009

Figure 2.1 Redrawn from Kaplan et al. 2000

Figure 2.2 Redrawn from Steward 1933, 326

Figure 2.3 Brink 2005; Friesen 2013; image from Brink 2005. Photo courtesy of Jack Brink.

Figure 2.4 From Swezey and Heizer 1977, 21. Copyright © Phoebe A. Hearst Museum of Anthropology and the Regents of the University of California, Photograph by P.E. Goddard (catalog no. 15-3301).

Figure 2.5 Redrawn from Mathew and Boyd 2011

Figure 2.6 Redrawn from Mathew and Boyd 2011

Figure 2.7 Redrawn from Mathew and Boyd 2011

Figure 2.8 Redrawn from Mathew and Boyd 2011

Figure 2.9 Redrawn from Mathew 2017

Figure 2.10 Redrawn from Sosis 2000

Figure 2.11 Redrawn from Sosis 2000

REFERENCES

CHAPTERS 1 AND 2

Alba, R., and Nee, V. 2003. *Remaking the American Mainstream: Assimilation and the New Immigration*. Harvard University Press, Cambridge, MA.

Aplin, K. P., Chesser, T., and Have, J. T. 2003. Evolutionary biology of the genus *Rattus*: Profile of an archetypal rodent pest. In *Rats, Mice and People: Rodent Biology and Management*, G. R. Singleton, L. A. Hinds, C. J. Krebs, and D. M. Spratt, eds., 487–498. Australian Centre for International Agricultural Research, Canberra.

Appiah, K. A. 2010. *The Honor Code: How Moral Revolutions Happen*. W. W. Norton, New York.

Arrow, K. 1962. The economic implications of learning by doing. *Review of Economic Studies*, 29, 155–173.

Axelrod, R., and Hamilton, W. D. 1981. The evolution of cooperation. *Science*, 211, 1390–1396.

Bamforth, D. P. 1994. Indigenous people, indigenous violence: Precontact warfare on the North American Great Plains. *Man*, 29, 95–115.

Barrett, C., Cosmides, L., and Tooby, J. 2007. The hominid entry into the cognitive niche. In *Evolution of Mind: Fundamental Questions and Controversies*, S. Gangestad and J. Simpson, eds., 241–248. Guilford Press, New York.

Barton, N. H., and Rouhani, S. 1993. Adaptation and the shifting balance. *Genetical Research*, 61, 57–74.

Basalla, G. 1988. *The Evolution of Technology*. Cambridge University Press, Cambridge, UK.

Beppu, A., and Griffiths, T. L. 2009. Iterated learning and the cultural ratchet. *Proceedings of the 31st Annual Conference of the Cognitive Science Society*. http://cocosci.berkeley.edu/tom/papers/ratchet1.pdf.

Binford, L. 2001. *Constructing Frames of Reference*. University of California Press, Berkeley.

Birch, S., and Okasha, S. 2014. Kin selection and its critics. *BioScience*, doi:10.1093/biosci/biu196.

Bisin, A., and Verdier, T. 2001. The economics of cultural transmission and the dynamics of preferences. *Journal of Economic Theory*, 97, 298–319.

Bliege-Bird, R., Bird, D. W., Codding, B. F., Parker, C. H., and Jones, J. H. 2009. The "fire stick farming" hypothesis: Australian Aboriginal foraging strategies, biodiversity, and anthropogenic fire mosaics. *Proceedings of the National Academy of Sciences (USA)*, 105, 14796–14801.

Boomsma, J. J., Beekman, M., Cornwallis, C. K., Griffin, A. S., Holman, L., Hughes, W. O. H., Keller, L., Oldroyd, B. P., and Ratnieks, F. L. W. 2011. Only full sibling families evolved eusociality. *Nature*, doi:10.1038/nature09832.

Bowles, S. 2006. Group competition, reproductive leveling, and the evolution of human altruism. *Science*, 314, 1569–1572.

Boyd, R., and Mathew, S. 2015.Third-party monitoring and sanctions aid the evolution of language. *Evolution and Human Behavior*, 36, 475–479.

Boyd, R., and Richerson, P. J. 1985. *Culture and the Evolutionary Process*. University of Chicago Press, Chicago.

———. 1987. The evolution of social learning: The effects of spatial and temporal variation. In *Social Learning: Psychological and Biological Perspectives*, T. R. Zentall and B. G. Galef, eds. Psychology Press, New York.

———. 1990. Group selection among alternative evolutionarily stable strategies. *Journal of Theoretical Biology*, 145, 331–342.

———. 1995. Why does culture increase human adaptability? *Ethology and Sociobiology*, 16, 125–143.

———. 1996. Why culture is common but cultural evolution is rare. *Proceedings of the British Academy*, 88, 73–93.

———. 2002. Group beneficial norms spread rapidly in a structured population. *Journal of Theoretical Biology*, 215, 287–296.

Boyd, R., Richerson, P. J., and Henrich, J. 2011a. The cultural niche: Why social learning is essential for human adaptation. *Proceedings of the National Academy of Sciences (USA)*, 108, 10918–10925.

———. 2011b. Rapid cultural adaptation can facilitate the evolution of large-scale cooperation. *Behavioral Ecology and Sociobiology*, 65, 431–444.

Boyd, R., Schonmann, R. H., and Vicente, R. 2014. Hunter-gatherer population structure and the evolution of contingent cooperation. *Evolution and Human Behavior*, 35, 219–227.

Boyer, P. 2001. *Religion Explained: The Evolutionary Origins of Religious Thought*. Basic Books, New York.

Brink, J. W. 2005. Inukshuk: Caribou drive lanes on southern Victoria Island, Nunavut, Canada. *Arctic Anthropology*, 42, 1–28.

———. 2008. *Imagining Head-Smashed-In: Aboriginal Buffalo Hunting on the Northern Great Plains*. Athabasca University Press, Edmonton, AB, Canada.

———. 2013. The Barnett site: A stone drive lane communal pronghorn trap on the Alberta Plains, Canada. *Quaternary International*, 297, 24–35.

Broesch, J. 2010. *Cultural Transmission and the Role of Evolutionary Forces: Empirical Evidence for Adaptive Biases in Learning and Cultural Transmission*. PhD diss., Emory University.

Cathcart, M. 2014. *Starvation in a Land of Plenty: Wills' Diary of the Fateful Burke and Wills Expedition*. National Library of Australia, Kindle ed.

Clarke, P. A. 2012. *Australian Plants as Aboriginal Tools*. Rosenberg Publishing, Kindle ed.

Clastres, P. 1972. The Guayaki. In *Hunters and Gatherers Today*, M. B. Bicchieri, ed. Holt, Rinehart and Winston, New York.

Clutton-Brock, T. 2009. Cooperation between non-kin in animal societies. *Nature*, 462, 51–57.

Collias, E., and Collias, N. 1964. The development of nest-building behavior in a weaverbird. *The Auk*, 81, 42–52.

———. 1973. Further studies on development of nest-building behavior in a weaverbird (*Ploceus cucullatus*). *Animal Behavior*, 21, 371–382.

Cosmides, L., and Tooby, J. 2006. Evolutionary psychology: A primer. Center for Evolutionary Psychology, University of California, Santa Barbara. http://www.cep.ucsb.edu/primer.html.

Couzin, I. D., and Franks, N. R. 2003. Self-organized lane formation and optimized traffic flow in army ants. *Proceedings of the Royal Society B*, 270, 139–146.

Cronk, L. 2002. From true Dorobo to Mukogodo Maasai: Contested ethnicity in Kenya. *Ethnology*, 41, 27–49.

Currie, T. E., Greenhill, S. J., Gray, R. D., Hasegawa, T., and Mace, R. 2010. Rise and fall of political complexity in island South-East Asia and the Pacific. *Nature*, 467, 801–804.

Dal Bó, P., and Fréchette, G. R. 2014. On the determinants of cooperation in infinitely repeated games: A survey. SSRN 2535963.

Darwent, J., and Darwent, C. M. 2014. Scales of violence across the North American Arctic. In *Violence and Warfare among Hunter-Gatherers*, M. W. Allen and T. L. Jones, eds., 149–167. Left Coast Press, Walnut Creek, CA.

Dawkins, R. 1986. *The Blind Watchmaker*. W. W. Norton, New York.

Dean, L. G., Val, G. L., Laland, K. N., Flynn, E., and Kendal, R. L. 2014. Human cumulative culture: A comparative perspective. *Biological Reviews*, 89, 284–301. doi: 0.1111/brv.12053.

Dehaene, S. 2009. *Reading and the Brain*. Penguin Viking, New York.

Denham, T., Fullagar, R., and Head, L. 2009. Plant exploitation on Sahul: From colonisation to the emergence of regional specialisation during the Holocene. *Quaternary International*, 202, 29–40.

Derex, M., and Boyd. R. 2015. The foundations of the human cultural niche. *Nature Communications*, doi:10.1038/ncomms9398.

Earl, J. W., and McCleary, B. V. 1994. Mystery of the poisoned expedition. *Nature*, 368, 683–684.

Enquist, M., Ghilanda, S., Jarrick, A., and Wachtmeister, C-A. 2008. Why does culture increase exponentially? *Theoretical Population Biology*, 74, 46–55.

Fitzpatrick, K. 2008. The Burke and Wills Expedition and the Royal Society of Victoria. *Historical Studies: Australia and New Zealand*, 10(40), 470–478. doi:10.1080/10314616308595251.

Franks, N. R. 1986. Teams in social insects: Group retrieval of prey by army ants (*Eciton burchelli*, Hymenoptera: Formicidae). *Behavioral Ecology and Sociobiology*, 18, 425–429.

Friesen, T. M. 2013. The impact of weapon technology on caribou drive system variability in the prehistoric Canadian Arctic. *Quaternary International*, 297, 13–23.

Fry, D. 2006. *The Human Potential for Peace: An Anthropological Challenge to Assumptions about War and Violence*. Oxford University Press, Oxford, UK.

Fudenberg, D., Rand, D. G., and Dreber, A. 2012. Slow to anger and fast to forgive: Cooperation in an uncertain world. *American Economic Review*, 102, 720–749.

Gat, A. 2015. Proving communal warfare among hunter-gatherers: The quasi-Rousseauan error. *Evolutionary Anthropology*, 24, 111–126.

Gavrilets, S. 1996. On phase three of the shifting-balance theory. *Evolution*, 50, 1034–1041.

Gergely, G., and Csibra, G. 2006. Sylvia's recipe: The role of imitation and pedagogy in the transmission of cultural knowledge. In *Roots of Human Sociality: Culture, Cognition, and Human Interaction*, N. J. Enfield and S. C. Levenson, eds., 229–255. Berg Publishers, Oxford, UK.

Gingrich, P. D. 1983. Rates of evolution: Effects of time and temporal scaling. *Science*, 222, 159–161.

Gould, J. L., and Gould, C. G. 2012. *Animal Architects: Building and the Evolution of Intelligence*. Basic Books, New York.

Greenhouse, S. 2014. More workers are claiming "wage theft." *New York Times*, August 31.

Guala, F. 2012. Reciprocity: Weak or strong? What punishment experiments do (and do not) demonstrate. *Behavioural and Brain Sciences*, 35, 1–59.

Gurven, M. 2004. Reciprocal altruism and food sharing decisions among Hiwi and Ache hunter-gatherers. *Behavioral Ecology and Sociobiology*, 56, 366–380.

Hailey, K., and Fessler, D. 2005. Nobody's watching? Subtle cues affect generosity in an anonymous economic game. *Evolution and Human Behavior*, 26, 245–256.

Hämäläinen, P. 2008. *The Comanche Empire*. Yale University Press, New Haven, CT.

Hansell, M. 2005. *Animal Architecture*. Oxford University Press, New York.

Hart, C. W. M., and Pilling, A. R. 1960. *The Tiwi of North Australia*. Holt, Rinehart and Winston, New York.

Henrich, J. 2004. Demography and cultural evolution: Why adaptive cultural processes produced maladaptive losses in Tasmania. *American Antiquity*, 69, 197–218.

———. 2006. Understanding cultural evolutionary models: A reply to Read's critique. *American Antiquity*, 71, 771–782.

———. 2009. The evolution of costly displays, cooperation and religion: Credibility enhancing displays and their implications for cultural evolution. *Evolution and Human Behavior*, 30, 244–260.

———. 2015. *The Secret of Our Success: How Culture Is Driving Human Evolution, Domesticating Our Species, and Making Us Smarter*. Princeton University Press, Princeton, NJ.

Henrich, J., and Boyd, R. 2001. Why people punish defectors: Weak conformist transmission can stabilize costly enforcement of norms in cooperative dilemmas. *Journal of Theoretical Biology*, 208, 79–89.

Henrich, J., Ensminger, J., McElreath, R., Barr, A., Barrett, C., Bolyanatz, A., Cardenas, J. C., Gurven, M., Gwako, E., Henrich, N., Lesorogo, C., Marlowe, F., Tracer, D., and Ziker, J. 2010. Markets, religion, community size, and the evolution of fairness and punishment. *Science*, 327, 1480–1484.

Henrich, J., and Gil-White, F. J. 2001. The evolution of prestige: Freely conferred deference as a mechanism for enhancing the benefits of cultural transmission. *Evolution and Human Behavior*, 22, 165–196.

Henrich, J., and Henrich, N. 2010. The evolution of cultural adaptations: Fijian food taboos protect against dangerous marine toxins. *Proceedings of the Royal Society B*, 277, 3715–3724.

Hill, K. 2002. Altruistic cooperation during foraging by the Ache, and the evolved human predisposition to cooperate. *Human Nature*, 13, 105–128.

Hill, K., Barton, M., and Hurtado, A. M. 2009. The origins of human uniqueness: The evolution of characters underlying behavioral modernity. *Evolutionary Anthropology*, 18, 187–200.

Hill, K. R., Walker, R. S., Božičević, M., Eder, J., Headland, T., Hewlett, B., Hurtado, A. M., Marlowe, F., Wiessner, P., and Wood, B. 2011. Co-residence patterns in hunter-gatherer societies show unique human social structure. *Science*, 331, 286–289.

Hill, K. R., Wood, B. M., Baggio, J., Hurtado, A. M., and Boyd, R. 2014. Hunter-gatherer inter-band interaction rates: Implications for cumulative culture. *PLoS One*, doi:10.1371/journal.pone.0102806.

Holden, C. J., and Mace, R. 2003. Spread of cattle led to the loss of matrilineal descent in Africa: A coevolutionary analysis. *Proceedings of the Royal Society B*, doi:10.1098/rspb.2003.2535.

Howitt, A. W., Burke, R. O., King, J., Mueller, F., Wills, W. J., Wright, W., Brahé, W., Ligon, C. W. V., MacAdam, J., Smith, J., and Archer, Mr. 1862. *Exploring Expedition from Victoria to the Gulf of Carpentaria, under the Command of Mr. Robert O'Hara Burke.* http://www.jstor.org/stable/1798420.

Iwaniuk, A. N., and Arnold, K. E. 2004. Is cooperative breeding associated with bigger brains? A comparative test in the Corvida (Passeriformes). *Ethology*, 110, 203–220.

Jacklyn, P. M. 1992. "Magnetic" termite mound surfaces are oriented to suit wind and shade conditions. *Oecologica*, 91, 385–395.

James, S. R., and Graziani, S. 1992. California Indian warfare. UC Berkeley Digital Assets. http://digitalassets.lib.berkeley.edu/anthpubs/ucb/text/arf 023-003.pdf.

Johnson, J. A. 1987. Dominance rank in juvenile olive baboons, Papio anubis: The influence of gender, size, maternal rank and orphaning. *Animal Behaviour*, 35, 1694–1708.

Johnson, S. 2010. *Where Good Ideas Come From: The Natural History of Innovation*. Riverhead Hardcover, New York.

Jordan, P. 2015. *Technology as Social Tradition: Cultural Transmission among Hunter Gatherers*. University of California Press, Berkeley.

Kaplan, H., Hill, K., Lancaster, J., and Hurtado, A. M. 2000. A theory of human life history evolution: Diet, intelligence, and longevity. *Evolutionary Anthropology*, 9, 156–185.

Kay, E. H., and Hoekstra, H. E. 2008. Rodents. *Current Biology*, 18, R408–R410.

Kelly, R. 1985. *The Nuer Conquest*. University of Michigan Press, Ann Arbor.

Klein, R. 2009. *The Human Career*. 3rd ed. University of Chicago Press, Chicago.

Kline, M., and Boyd, R. 2010. Population size predicts technological complexity in Oceania. *Proceedings of the Royal Society B*, 277, 2559–2564.

Kroeber, A. L. (1925) 1976. *The Handbook of California Indians*. Dover, New York.

Lane, N. 2009. *Life Ascending*. Profile Books, London.

Langergraber, K., Schubert, G., Rowney, C., Wrangham, R., Zommers, Z., and Vigilant, L. 2011. Genetic differentiation and the evolution of cooperation in chimpanzees and humans. *Proceedings of the Royal Society B*, doi:10.1098/rspb.2010.2592.

LeBlanc, S. 2014. Forager warfare and our evolutionary past. In *Violence and Warfare among Hunter-Gatherers*, M. W. Allen and T. L. Jones, eds. 149–167. Left Coast Press, Walnut Creek, CA.

Lehmann, L., and Keller, L. 2006. The evolution of cooperation and altruism: A general framework and a classification of models. *Journal of Evolutionary Biology*, 19, 1365–1376.

Lehmann, L., Perrin, N., and Rousset, F. 2006. Population demography and the evolution of helping behaviors. *Evolution*, 60, 1137–1151.

Leimar, O. 1997. Repeated games: A state space approach. *Journal of Theoretical Biology*, 184, 471–498.

Lewis, J. 2014. Egalitarian social organization: The case of the Mbendjele BaYaka. In *Hunter-Gatherers of the Congo Basin*, B. S. Hewlett, ed., Kindle location 5800. Transaction Publishers, Kindle ed.

Lukas, D., and Clutton-Brock, T. 2012a. Cooperative breeding and monogamy in mammalian societies. *Proceedings of the Royal Society B*, 279, 2151–2156.

———. 2012b. Life histories and the evolution of cooperative breeding in mammals. *Proceedings of the Royal Society B*, 279, 4065–4070.

Manson, J., and Wrangham, R. 1991. Intergroup aggression in chimpanzees and humans. *Current Anthropology*, 32, 369–377.

Marlowe, F. 2009. Hadza cooperation: Second party punishment, yes; third party punishment, no. *Human Nature*, 20, 417–430.

Martin, P. 2005. Migrants in the global labor market. Global Commission on International Migration Working Paper. http://www.iom.int/ jahia/webdav/site/myjahiasite/shared/shared/mainsite/policy_and_research/gcim/tp/TP1.pdf.

Maskin, E., Harrison, L., and Yasin, Y. 2015. Culture, cooperation, and repeated games. In *Culture Matters in Russia—and Everywhere*, L. Harrison and E. Yasin, eds. Lexington Press, Lanham, MD.

Mathew, S. 2017. How the second-order free rider problem is solved in a small-scale society. *American Economic Review*, 107, 1–4. https://doi.org/10.1257/aer.p20171090.

Mathew, S., and Boyd, R. 2011. Punishment sustains large-scale cooperation in prestate warfare. *Proceedings of the National Academy of Sciences (USA)*, 108, 11375–11380.

———. 2014. The cost of cowardice. *Evolution and Human Behavior*, 35, 58–64.

———. Forthcoming. Adjudication and perception errors.

Mathew, S., Boyd, R., and van Veelen, M. 2013. Human cooperation among kin and close associates may require enforcement of norms by third parties. In *Cultural Evolution*, P. J. Richerson and M. Christiansen, eds. Strüngmann Forum Report 12, J. Lupp, series ed. MIT Press, Cambridge, MA.

Mathew, S., and Perreault, C. 2015. Behavioural variation in 172 small-scale societies indicates that social learning is the main mode of human adaptation. *Proceedings of the Royal Society B*, doi:10.1098/ rspb.2015.0061.

McElreath, R., Bell, A. V., Efferson, C., Lubell, M., Richerson, P. J., and Waring, T. 2008. Beyond existence and aiming outside the laboratory: Estimating frequency-dependent and payoff-biased social learning strategies. *Philosophical Transactions of the Royal Society B*, 363, 3515–3528.

McElreath, R., and Boyd, R. 2007. *Mathematical Models of Social Evolution: A Guide for the Perplexed.* University of Chicago Press, Chicago.

Meggitt, M. J. 1965. *Desert People: A Study of the Walbiri Aborigines of Central Australia.* University of Chicago Press, Chicago.

Moorehead, A. 1963. *Cooper's Creek.* Dell, New York.

Morgan, J. (1852) 2002. *The Life and Adventures of William Buckley.* Text Publishing, Melbourne, Australia.

Morgan, T. H., Uomini, N., Rendell, L. E., Chouinard-Thuly, L., Street, S., and Lewis, H. 2015. Experimental evidence for the co-evolution of hominin tool-making teaching and language. *Nature Communications* (January), 6. doi:10.1038/ncomms7029.

Nowak, M., Page, K., and Sigmund, K. 2000. Fairness versus reason in the ultimatum game. *Science*, 289, 1773–1775.

Nowak, M., and Sigmund, K. 1993. A strategy of win-stay, lose-shift that outperforms tit-for-tat in the Prisoner's Dilemma game. *Nature*, 364, 56–58.

Nowak, R. M. 1999. *Walker's Mammals of the World.* Johns Hopkins University Press, Baltimore.

Neiman, F. D. 1995. Stylistic variation in evolutionary perspective: Inferences from decorative diversity and interassemblage distance in Illinois Woodland ceramic assemblages. *American Antiquity*, 60, 7–36.

O'Shea, J., Lemke, A. K., and Reynolds, R. G. 2013. "Nobody knows the way of the caribou": *Rangifer* hunting at 45° north latitude. *Quaternary International*, 297, 36–44.

Panchanathan, K., and Boyd, R. 2004. Indirect reciprocity can stabilize cooperation without the second-order free rider problem. *Nature*, 432, 499–502.

Perreault, C. 2012. The pace of cultural evolution. *PLoS ONE*, doi:10.1371/journal.pone.0045150.

Perreault, C., Moya, C., and Boyd, R. 2012. A Bayesian approach to the evolution of social learning. *Evolution and Human Behavior*, 33, 449–459.

Petroski, H. 1992a. *The Evolution of Useful Things: How Everyday Artifacts—From Forks and Pins to Paper Clips and Zippers—Came to Be as They Are.* Vintage Books, New York.

———. 1992b. *The Pencil.* Knopf, New York.

Pinker, S. 2010. The cognitive niche: Coevolution of intelligence, sociality, and language. *Proceedings of the National Academy of Sciences (USA)*, 107, 8993–8999.

Powell, A., Shennan, S., and Thomas, M. G. 2009. Late Pleistocene demography and the appearance of modern human behavior. *Science*, 324, 1298–1301.

Raihani, N., Thornton, A., and Bshary, R. 2012. Punishment and cooperation in nature. *Trends in Ecology and Evolution*, 27, 288–295.

Richerson, P. J., Bell, A., Demps, K., Frost, K., Hillis, V., Mathew, S., Newton, E., Narr, N., Newson, L., Ross, C., Smaldino, P., Waring, T., and Zefferman, M. 2016. Cultural group selection plays an essential role in explaining human cooperation: A sketch of the evidence. *Behavioural and Brain Sciences*, 39, 1–68. doi:10.1017/S0140525X1400106X, e30.

Richerson, P. J., and Boyd, R. 2005. *Not by Genes Alone*. University of Chicago Press, Chicago.

Romer, P. 1993. Endogenous technological change. *Journal of Political Economy*, 98, 71–102.

Scherjon, F., Bakels, C., MacDonald, K., and Roebroeks, W. 2015. Burning the land: An ethnographic study of off-site fire use by current and historically documented foragers and implications for the interpretation of past fire practices in the landscape. *Current Anthropology*, 56, 299–326.

Schino, G., and Aureli, F. 2009. Reciprocal altruism in primates: Partner choice, cognition, and emotions. *Advances in the Study of Behavior*, 39, 45–69.

Schonmann, R. H., and Boyd, R. 2016. A simple rule for the evolution of contingent cooperation in large groups. *Philosophical Transactions of the Royal Society B*, dx.doi.org/10.1098/rstb.2015.0099.

Seabright, P. 2010. *In the Company of Strangers*. Princeton University Press, Princeton, NJ.

Sharif, A., Norenzayan, A., and Henrich, J. 2010. The birth of high gods: How the cultural evolution of supernatural policing influenced the evolution of complex cooperative human societies paving the way for civilization. In *Evolution, Culture and the Human Mind*, M. Schaller, A. Norenzayan, S. J. Heine, T. Yamagishi, and T. Kameda, eds., 119–136. Taylor and Francis, New York.

Shennan, S. 2001. Demography and cultural innovation: A model and its implications for the emergence of modern human culture. *Cambridge Archaeological Journal*, 11, 5–16.

Smith, B. D. 2011. General patterns of niche construction and the management of "wild" plant and animal resources by small-scale pre-industrial societies. *Philosophical Transactions of the Royal Society B*, 366, 836–848.

Smith, M. W. 1938. The war complex of the Plains Indians. *Proceedings of the American Philosophical Society*, 78, 425–464.

Soltis, J., Boyd, R., and Richerson, P. J. 1995. Can group functional behaviors evolve by cultural group selection? An empirical test. *Current Anthropology*, 36, 473–494.

Sosis, R. 2000. Religion and intragroup cooperation: Preliminary results of a comparative analysis of utopian communities. *Cross Cultural Research*, 34, 70–88.

Stark, R. 1997. *The Rise of Christianity: How the Obscure, Marginal Jesus Movement Became the Dominant Religious Force in the Western World in a Few Centuries*. HarperCollins, San Francisco.

Steinmetz, P. H. R., Kraus, J. E. M., Larroux, C., Jörg, U., Hammel, J. U., Amon-Hassenzahl, A., Houliston, E., Wörheide, G., Nickel, M., Degnan, B. M., and Technau, U. 2012. Independent evolution of striated muscles in cnidarians and bilaterians. *Nature*, 487, 231–236.

Sterelny, K. 2012. *The Evolved Apprentice: How Evolution Made Humans Unique*. MIT Press, Cambridge, MA.

Stevens, J. R., and Stephens, D. W. 2003. The economic basis of cooperation: Tradeoffs between selfishness and generosity. *Behavioral Ecology*, 15, 255–261.

Steward, J. 1933. Ethnography of the Owens Valley Paiute. *University of California Publications in American Archaeology Ethnology*, 33, 233–350.

Sugiyama, L. S. 2003. Illness, injury, and disability among Shiwiar forager-horticulturalists: Implications of health-risk buffering for the evolution of human life history. *American Journal of Physical Anthropology*, 123, 371–389.

Sutton, M. Q. 2014. Warfare and expansion: An ethnohistoric perspective on the Numic spread. In *Violence and Warfare among Hunter-Gatherers*, M. W. Allen and T. L. Jones, eds., 149–167. Left Coast Press, Walnut Creek, CA.

Swezey, S. L., and Heizer, R. H. 1977. Ritual management of salmonid fish resources in California. *Journal of California Anthropology*, 4, 429.

Tebbich, S., Taborsky, M., Fessl, B., and Blomqvist, D. 2001. Do woodpecker finches acquire tool-use by social learning? *Proceedings of the Royal Society B*, 268, 2189–2193.

Tennie, C., Call, J., and Tomasello, M. 2009. Ratcheting up the ratchet: On the evolution of cumulative culture. *Philosophical Transactions of the Royal Society B*, 364, 2405–2415.

Tooby, J., and DeVore, I. 1987. The cognitive niche. In *Primate Models of Hominid Behavior*, W. Kinzey, ed., 183–237. SUNY Press, New York.

Trivers, R. L. 1971. The evolution of reciprocal altruism. *Quarterly Review of Biology*, 46, 35–57.

Wadley, L., Hodgskiss, T., and Grant, M. 2009. Implications for complex cognition from the hafting of tools with compound adhesives in the Middle Stone Age, South Africa. *Proceedings of the National Academy of Sciences (USA)*, 106, 9590–9594.

West, S. A., El Moden, C., and Gardner, A. 2011. Sixteen common misconceptions about the evolution of cooperation in humans. *Evolution and Human Behavior*, 32, 231–262.

Whiten, A., Horne, V., and de Waal, F. B. M. 2005. Conformity to cultural norms of tool use in chimpanzees. *Nature*, 437, 737–740.

Wilke, P. 2013. The Whisky Flat pronghorn trap complex, Mineral County, Nevada, western United States: Preliminary report. *Quaternary International*, 297, 79–92.

Wood, B., and Gilby, I. Forthcoming. Hunting and meat sharing by chimpanzees, humans, and our common ancestor. In *Chimpanzees and Human Evolution*, M. Muller, R. Wrangham, and D. Pilbeam, eds. Harvard University Press, Cambridge, MA.

Young, P. 2001. *Individual Strategy and Social Structure: An Evolutionary Theory of Institutions*. Princeton University Press, Princeton, NJ.

Zefferman, M. R., and Mathew, S. 2015. An evolutionary theory of large-scale human warfare: Group-structured cultural selection. *Evolutionary Anthropology*, 24, 50–61.

CHAPTER 3

Boyd, R., and Richerson, P. J. 1993. Rationality, imitation, and tradition. In *Nonlinear Dynamics and Evolutionary Economics*. R. Day and P. Chen, eds., 131–149. Oxford University Press, New York.

Burke, E. (1790) 1910. *Reflections on the Revolution in France*. J. M. Dent and Sons, London.

Cavalli-Sforza, L. L., and Feldman, M. W. 1981. *Cultural Transmission and Evolution: A Quantitative Approach*. Princeton University Press, Princeton, NJ.

Hayek, F. A., 1952. *The Counter-Revolution of Science*. Free Press, Glencoe, IL.

———. 1984a. Dr. Bernard Mandeville. In *The Essence of Hayek*, C. Nishiyama and K. R. Leube, eds., 176–194. Hoover Institution Press, Stanford, CA.

———. 1984b. *The Essence of Hayek*. C. Nishiyama and K. R. Leube, eds. Hoover Institution Press, Stanford, CA.

Lescak, E. A., Bassham, S. L., Catchen, J., Gelmond, O., Sherbick, M. L., von Hippel, F. A., and Cresko, W. A. 2015. Evolution of stickleback in 50 years on earthquake-uplifted islands. *Proceedings of the National Academy of Sciences (USA)*, 112, E7204–E7212.

Orr, H. A. 2009. Darwin and Darwinism: The (alleged) social implications of *The Origin of Species*. *Genetics*, 183, 767–772.

———. 2015. The biology of being good to others (review of David Sloan Wilson's *Does Altruism Exist?*) *New York Review of Books* (March), 1927–1929.

Perreault, C., Moya, C., and Boyd, R. 2012. A Bayesian approach to the evolution of social learning. *Evolution and Human Behavior*, 33, 449–459.

Price, G. R. 1970. Selection and covariance. *Nature*, 227(5257), 520–521.

Price, G. R. 1972. Extension of covariance selection mathematics. *Annals of Human Genetics*, 35, 485–490.

Secord, J. A., ed. 2008. *Charles Darwin: Evolutionary Writings*. Oxford University Press, Oxford, UK.

Subiaul, F. 2007. The imitation faculty in monkeys: Evaluating its features, distribution and evolution. *Journal of Anthropological Sciences*, 85, 35–62.

Tooby, J., and DeVore, I. 1987. The cognitive niche. In *Primate Models of Hominid Behavior*, W. Kinsey, ed., 183–237. SUNY Press, New York.

CHAPTER 4

Alvard, M., and Nolin, D. 2002. Rousseau's whale hunt? Coordination among big game hunters. *Current Anthropology*, 43(4), 533–559.

Edgerton, R. B. 1992. *Sick Societies: Challenging the Myth of Primitive Harmony*. Free Press, New York.

Ewald, P. W. 1994. *Evolution of Infectious Disease*. Oxford University Press, Oxford, UK.

Harris, M. 1985. *Good to Eat: Riddles of Food and Culture*. Simon and Schuster, New York.

Henrich, J. 2004. Demography and cultural evolution: Why adaptive cultural processes produced maladaptive losses in Tasmania. *American Antiquity*, 69(2), 197–221.

———. 2016. *The Secret of Our Success: How Culture Is Driving Human Evolution, Domesticating Our Species and Making Us Smarter*. Princeton University Press, Princeton, NJ.

Henrich, J., and Henrich, N. 2010. The evolution of cultural adaptations: Fijian food taboos protect against dangerous marine toxins. *Proceedings of the Royal Society B*, 277, 3715–3724.

Hewlett, B., Fouts, H., Boyette, A., and Hewlett, B. 2011. Social learning among Congo Basin hunter-gatherers. *Philosophical Transactions of the Royal Society of London B*, 366(1567), 1168–1178.

Heyes, C. 2012. Grist and mills: On the cultural origins of cultural learning. *Philosophical Transactions of the Royal Society of London B*, 367(1599), 2181–2191.

Hiscock, P. 2014. Learning in lithic landscapes: A reconsideration of the hominid "toolmaking" niche. *Biological Theory*, 9(1), 27–41.

Liebenberg, L. 1990. *The Art of Tracking and the Origin of Science*. David Philip, Claremount, South Africa.

———. 2008. The relevance of persistence hunting to human evolution. *Journal of Human Evolution*, 55(6), 1156–1159.

———. 2013. *The Origin of Science: On the Evolutionary Roots of Science and Its Implications for Self-Education and Citizen Science*. CyberTracker, Capetown, South Africa.

McBrearty, S., and Brooks, A. 2000. The revolution that wasn't: A new interpretation of the origin of modern human behavior. *Journal of Human Evolution*, 39(5), 453–563.

Morin, O. 2016. *How Traditions Live and Die*. Oxford University Press, Oxford, UK.

Nichols, S. 2004. *Sentimental Rules: On the Natural Foundations of Moral Judgment*. Oxford University Press, New York.

Odling-Smee, J., Laland, K., and Feldman, M. 2003. *Niche Construction: The Neglected Process in Evolution*. Princeton University Press, Princeton, NJ.

Powell, A., Shennan, S., and Thomas, M. 2009. Late Pleistocene demography and the appearance of modern human behavior. *Science*, 324 (June 5), 298–1301.

Richerson, P., and Boyd, R. 2013. Rethinking paleoanthropology: A world queerer than we had supposed. In *The Evolution of Mind, Brain and Culture*, G. Hatfield and H. Pittman, eds., 263–302. University of Pennsylvania Press, Philadelphia.

Shaw-Williams, K. 2014. The social trackways theory of the evolution of human cognition. *Biological Theory*, 9(1), 16–26.

Sterelny, K. 2012. *The Evolved Apprentice*. MIT Press, Cambridge, MA.

Stout, D. 2002. Skill and cognition in stone tool production: An ethnographic case study from Irian Jaya. *Current Anthropology*, 43(5), 693–722.

CHAPTER 5

Bauer, M., Cassar, A., Chytilova, J., and Henrich, J. 2014. War's enduring effects on the development of egalitarian motivations and in-group biases. *Psychological Science*, 25(1), 47–57.

Boyd, R., and Richerson, P. J. 1985. *Culture and the Evolutionary Process*. University of Chicago Press, Chicago.

Dawkins, R. 1976. *The Selfish Gene*. Oxford University Press, Oxford, UK.

Gneezy, A., and Fessler, D. M. T. 2012. Conflict, sticks and carrots: War increases prosocial punishments and rewards. *Proceedings of the Royal Society B*, 279(1727), 219–223.

Mace, R. 2014. Human behavioral ecology and its evil twin. *Behavioral Ecology*, 25(3), 443–449.

Silva, A. S., and Mace, R. 2014. Cooperation and conflict: Field experiments in Northern Ireland. *Proceedings of the Royal Society B*, 281(1792).

———. 2015. Inter-group conflict and cooperation: Field experiments before, during and after sectarian riots in Northern Ireland. *Frontiers in Psychology*, 6.

Ting, J., Zheng, X.-D., He, Q.-Q., Wu, J.-J., Mace, R., and Tao, Y. 2016. Kinship as a frequency-dependent strategy. *Royal Society Open Science* 3(2).

CHAPTER 6

Booker, C. 2006. *The Seven Basic Plots: Why We Tell Stories*. Bloomsbury, London.

Boyd, R., and Richerson, P. J. 1985. *Culture and the Evolutionary Process*. University of Chicago Press, Chicago.

Campbell, J. 1949. *The Hero with a Thousand Faces*. Princeton University Press, Princeton, NJ.

Graça da Silva, S., and Tehrani, J. J. 2016. Comparative phylogenetic analyses uncover the ancient roots of Indo-European folktales. *Royal Society Open Science*, 3, 150645. http://dx.doi.org/10.1098/rsos.150645.

Henrich, J. 2016. *The Secret of Our Success: How Culture Is Driving Human Evolution, Domesticating Our Species and Making Us Smarter*. Princeton University Press, Princeton, NJ.

Mathew, S., and Boyd, R. 2011. Punishment sustains large-scale cooperation in prestate warfare. *Proceedings of the National Academy of Sciences (USA)*, 108, 11375–11380.

CHAPTER 7

Barrett, H. C., and Broesch, J. 2012. Prepared social learning about dangerous animals in children. *Evolution and Human Behavior*, 33, 499–508.

Bisin, A., and Verdier, T. 2001. The economics of cultural transmission and the dynamics of preferences. *Journal of Economic Theory*, 97, 298–319.

Boyd, R., and Richerson, P. J. 1985. *Culture and the Evolutionary Process*. University of Chicago Press, Chicago.

Carey, S. 2011. *The Origin of Concepts*. Oxford University Press, Oxford, UK. Kindle ed.

Claidière, N., and Sperber, D. 2007. The role of attraction in cultural evolution. *Journal of Cognition and Culture*, 7, 89–111.

Claidière, N., and Whiten, A. 2012. Integrating the study of conformity and culture in humans and nonhuman animals. *Psychological Bulletin*, 138, 126–145.

Dehaene, S. 2009. *Reading in the Brain*. Viking, New York.

Derex, M., and Boyd, R. Forthcoming. What do people learn when they study socially?

Edgerton, R. 1985. *Rules, Exceptions and Social Order*. University of California Press, Berkeley.

———. 1992. *Sick Societies: Challenging the Myth of Primitive Harmony*. Free Press, New York.

Feil, D. K. 1987. *The Evolution of Highland Papua New Guinea Societies*. Cambridge University Press, Cambridge.

Henrich, J. 2009. The evolution of costly displays, cooperation and religion: Credibility enhancing displays and their implications for cultural evolution. *Evolution and Human Behavior*, 30, 244–260.

———. 2015. *The Secret of Our Success: How Culture Is Driving Human Evolution, Domesticating Our Species, and Making Us Smarter.* Princeton University Press, Princeton, NJ.

Henrich, J., and Boyd, R. 1998. The evolution of conformist transmission and the emergence of between-group differences. *Evolution and Human Behavior*, 19, 215–242.

Henrich, J., and McElreath, R. 2003. The evolution of cultural evolution. *Evolutionary Anthropology*, 12, 123–135.

Hill, K., and Hurtado, A. M. 2009. Cooperative breeding in South American hunter-gatherers. *Proceedings of the Royal Society B*, 276, 3863–3870.

Jaeggi, A., and Gurven, M. 2013. Natural cooperators: Food sharing in humans and other primates. *Evolutionary Anthropology*, 22, 186–195.

Kaplan, H. S., Lancaster, J. B., Johnson, S. E., and Bock, J. 1995. Does observed fertility maximize fitness among New Mexican men? *Human Nature*, 6, 325–360.

Kimbel, W. H., and Villmoare, B. 2016. From *Australopithecus* to *Homo*: The transition that wasn't. *Philosophical Transactions of the Royal Society B*, 371, 20150248. http://dx.doi.org/10.1098/rstb.2015.0248.

Kline, M. A., Boyd, R., and Henrich, J. 2013. Teaching and the life history of cultural transmission in Fijian villages. *Human Nature*, 24, 351–374.

Knauft, B. 1985. *Good Company and Violence: Sorcery and Social Action in a Lowland New Guinea Society.* University of California Press, Berkeley.

Mikhail, J. 2007. Universal moral grammar: Theory, evidence and the future. *Trends in Cognitive Science*, 11, 143–152.

Oliver, S. C. 1962. *Ecology and Cultural Continuity as Contributing Factors in the Social Organization of the Plains Indians.* University of California Publications in American Archaeology and Ethnology 48, Berkeley.

Perreault, C., Moya, C., and Boyd, R. 2012. A Bayesian approach to the evolution of social learning. *Evolution and Human Behavior*, 33, 449–459.

Pica, P., Lemer, C., Izard, V., and Dehaene, S. 2004. Exact and approximate arithmetic in an Amazonian indigene group. *Science*, 306, 499–503.

Richerson, P. J., Baldini, R., Bell, A. V., Demps, K., Frosta, K., Hillis, V., Mathew, S., Newton, E. K., Naara, E., Newson, L., Ross, C., Smaldino, P. E., Waringa, T. M., and Zefferman, M. 2016. Cultural group selection plays an essential role in explaining human cooperation: A sketch of the evidence. *Behavioural and Brain Sciences*, 39, 1–68. doi:10.1017/S0140525X1400106X, e30.

Richerson, P. J., and Boyd, R. 2005. *Not by Genes Alone*. University of Chicago Press, Chicago.

Scott-Philips, T. 2014. *Speaking Our Minds: Why Human Communication Is Different, and How Language Evolved to Make It Special*. Palgrave MacMillan, Basingstoke, UK. Kindle ed.

Seabright, P. A. 2010. *In the Company of Strangers: A Natural History of Economic Life*. Rev. ed. Princeton University Press, Princeton, NJ.

Vaesen, K. 2012. The cognitive bases of human tool use. *Behavioural and Brain Sciences*, 35, 203–218.

Wakano, J., and Aoki, K. 2007. Do social learning and conformist bias coevolve? Henrich and Boyd revisited. *Theoretical Population Biology*, 72, 504–512.

Wertz, A. E., and Wynn, K. 2014. Selective social learning of plant edibility in 6- and 18-month-old infants. *Psychological Science*, 25, 874–882.

Wiessner, P. W., Tumu, A., and Pupu, N. 1998. *Historical Vines: Enga Networks of Exchange, Ritual, and Warfare in Papua New Guinea*. Smithsonian Institution Press, Washington, DC.

Zipes, J. 2013. *The Irresistible Fairy Tale: The Social and Cultural History of a Genre*. Princeton University Press, Princeton, NJ.

Contributors

Robert Boyd is Origins Professor in the School of Human Evolution and Social Change at Arizona State University. His research focuses on incorporating cultural transmission into the Darwinian theory of evolution, and on the evolution of social behavior, especially reciprocity and collective action. Some of this work is described in *Culture and the Evolutionary Process*, *Not by Genes Alone*, and *The Origin and Evolution of Cultures*, all co-authored with Peter Richerson. He has also written an overview of human evolution, *How Humans Evolved*, with collaboration with Joan Silk.

Ruth Mace is Professor of Evolutionary Anthropology, University College London. She works in many areas of evolutionary anthropology, including the behavioral ecology of kinship and social organization in Africa and Asia, most recently in China. She was elected as a Fellow of the British Academy in 2008, and has held visiting professorships at the Chinese Academy of Sciences, Beijing, and at Lanzhou University, PRC. Ruth is currently President of the European Human Behaviour and Evolution Association (EHBEA).

Stephen Macedo is the Laurance S. Rockefeller Professor of Politics and the University Center for Human Values at Princeton University, where he also chairs the Tanner Lecture Committee. He is past director of the University Center for Human Values (2001–2009), and the founding director of Princeton's Program in Law and Public Affairs (1999–2001). He is author, most recently, of *Just Married: Same-Sex Couples, Monogamy, and the Future of Marriage* (Princeton University Press, 2015).

H. ALLEN ORR is University Professor and Shirley Cox Kearns Professor of Biology at the University of Rochester. He is an evolutionary geneticist whose work focuses on speciation, extinction, and adaptation. He is the author of *Speciation* (with Jerry A. Coyne). He is also a frequent book reviewer and has written for *The New York Review of Books* and *The New Yorker*.

PAUL SEABRIGHT is Professor of Economics at the Toulouse School of Economics and Director of the Institute for Advanced Study in Toulouse. He is author, among other works, of *The Company of Strangers: A Natural History of Economic Life* (Princeton, 2nd edition, 2010).

KIM STERELNY is an Australian philosopher and professor of philosophy in the Research School of Social Sciences at Australian National University and Victoria University of Wellington. He has written extensively on the philosophy of science, especially biology. His books include *Thought in a Hostile World* and *The Evolved Apprentice*. Until recently, he edited *Biology and Philosophy*. He is also a member of the Australian Academy of the Humanities.

Index

Page numbers in italics refer to figures.

The University Center for Human Values Series
Stephen Macedo, Editor

Multiculturalism and "The Politics of Recognition"
by Charles Taylor

A Matter of Interpretation: Federal Courts and the Law
by Antonin Scalia

Freedom of Association edited by Amy Gutmann

Work and Welfare by Robert M. Solow

The Lives of Animals by J. M. Coetzee

Truth v. Justice: The Morality of Truth Commissions edited
by Robert I. Rotberg and Dennis Thompson

Goodness and Advice by Judith Jarvis Thomson

Human Rights as Politics and Idolatry by Michael Ignatieff

Democracy, Culture, and the Voice of Poetry by Robert Pinsky

Primates and Philosophers: How Morality Evolved
by Frans de Waal

Striking First: Preemption and Prevention in International Conflict
by Michael W. Doyle

Meaning in Life and Why It Matters by Susan Wolf

The Limits of Constitutional Democracy edited
by Jeffrey K. Tulis and Stephen Macedo

Foragers, Farmers, and Fossil Fuels: How Human Values Evolve
by Ian Morris

*Private Government: How Employers Rule Our Lives
(and Why We Don't Talk about It)*
by Elizabeth Anderson

A Different Kind of Animal: How Culture Transformed Our Species
by Robert Boyd